DECISION MAKING *for the* ENVIRONMENT

Social and Behavioral Science Research Priorities

Panel on Social and Behavioral Science Research Priorities for Environmental Decision Making

Garry D. Brewer and Paul C. Stern, editors

Committee on the Human Dimensions of Global Change

Center for Governance, Economics, and International Studies

Division of Behavioral and Social Sciences and Education

NATIONAL RESEARCH COUNCIL
OF THE NATIONAL ACADEMIES

THE NATIONAL ACADEMIES PRESS
Washington, D.C.
www.nap.edu

THE NATIONAL ACADEMIES PRESS 500 Fifth Street, N.W. Washington, DC 20001

NOTICE: The project that is the subject of this report was approved by the Governing Board of the National Research Council, whose members are drawn from the councils of the National Academy of Sciences, the National Academy of Engineering, and the Institute of Medicine. The members of the committee responsible for the report were chosen for their special competences and with regard for appropriate balance.

This study was supported by Contract/Grant No. X-83060501 between the National Academy of Sciences and the U.S. Environmental Protection Agency. Any opinions, findings, conclusions, or recommendations expressed in this publication are those of the author(s) and do not necessarily reflect the views of the organizations or agencies that provided support for the project.

Library of Congress Cataloging-in-Publication Data

Decision making for the environment : social and behavioral science research priorities / Panel on Social and Behavioral Science Research Priorities for Environmental Decision Making, Committee on the Human Dimensions of Global Change [and] Center for Governance, Economics, and International Studies, Division of Behavioral and Social Sciences and Education ; Garry D. Brewer and Paul C. Stern, editors.
p. cm.
Includes bibliographical references.
ISBN 0-309-09540-9 (pbk.) — ISBN 0-309-54772-5 (pdf) 1. Environmental policy—Decision making. 2. Environmental policy—Research. 3. Research—Environmental aspects. I. Brewer, Garry D. II. Stern, Paul C., 1944- III. National Research Council (U.S.). Panel on Social and Behavioral Science Research Priorities for Environmental Decision Making. IV. National Research Council (U.S.). Center for Governance, Economics, and International Studies.
GE170.D4435 2005
333.72—dc22 2005009679

Additional copies of this report are available from National Academies Press, 500 Fifth Street, N.W., Lockbox 285, Washington, DC 20055; (800) 624-6242 or (202) 334-3313 (in the Washington metropolitan area); Internet, http://www.nap.edu

Printed in the United States of America

Suggested citation: National Research Council. (2005). *Decision Making for the Environment: Social and Behavioral Science Research Priorities*. Panel on Social and Behavioral Science Research Priorities for Environmental Decision Making. G.D. Brewer and P.C. Stern, editors. Committee on the Human Dimensions of Global Change, Division of Behavioral and Social Sciences and Education. Washington, DC: The National Academies Press.

THE NATIONAL ACADEMIES

Advisers to the Nation on Science, Engineering, and Medicine

The **National Academy of Sciences** is a private, nonprofit, self-perpetuating society of distinguished scholars engaged in scientific and engineering research, dedicated to the furtherance of science and technology and to their use for the general welfare. Upon the authority of the charter granted to it by the Congress in 1863, the Academy has a mandate that requires it to advise the federal government on scientific and technical matters. Dr. Bruce M. Alberts is president of the National Academy of Sciences.

The **National Academy of Engineering** was established in 1964, under the charter of the National Academy of Sciences, as a parallel organization of outstanding engineers. It is autonomous in its administration and in the selection of its members, sharing with the National Academy of Sciences the responsibility for advising the federal government. The National Academy of Engineering also sponsors engineering programs aimed at meeting national needs, encourages education and research, and recognizes the superior achievements of engineers. Dr. Wm. A. Wulf is president of the National Academy of Engineering.

The **Institute of Medicine** was established in 1970 by the National Academy of Sciences to secure the services of eminent members of appropriate professions in the examination of policy matters pertaining to the health of the public. The Institute acts under the responsibility given to the National Academy of Sciences by its congressional charter to be an adviser to the federal government and, upon its own initiative, to identify issues of medical care, research, and education. Dr. Harvey V. Fineberg is president of the Institute of Medicine.

The **National Research Council** was organized by the National Academy of Sciences in 1916 to associate the broad community of science and technology with the Academy's purposes of furthering knowledge and advising the federal government. Functioning in accordance with general policies determined by the Academy, the Council has become the principal operating agency of both the National Academy of Sciences and the National Academy of Engineering in providing services to the government, the public, and the scientific and engineering communities. The Council is administered jointly by both Academies and the Institute of Medicine. Dr. Bruce M. Alberts and Dr. Wm. A. Wulf are chair and vice chair, respectively, of the National Research Council.

www.national-academies.org

PANEL ON SOCIAL AND BEHAVIORAL SCIENCE RESEARCH PRIORITIES FOR ENVIRONMENTAL DECISION MAKING

Garry D. Brewer *(Chair)*, School of Management, Yale University
Braden R. Allenby, Ira A. Fulton School of Engineering, Arizona State University
Richard N. Andrews, Institute for Environmental Studies, University of North Carolina, Chapel Hill
Susan L. Cutter, Department of Geography, University of South Carolina
J. Clarence Davies, Resources for the Future, Washington, DC
Loren Lutzenhiser, College of Urban Studies and Planning, Portland State University
Bonnie J. McCay, Department of Human Ecology, Cook College of Rutgers University
Timothy McDaniels, School of Community and Regional Planning, University of British Columbia
Jennfier Nash, Regulatory Policy Program, John F. Kennedy School of Government
Steven W. Percy, University of Michigan Business School, Ann Arbor
David Skole, Department of Geography, Michigan State University, East Lansing
Neil Weinstein,* Department of Human Ecology, Cook College of Rutgers University

Paul C. Stern, *Study Director*
Deborah Johnson, *Senior Project Assistant*

*Resigned June 12, 2003

COMMITTEE ON THE HUMAN DIMENSIONS OF GLOBAL CHANGE

Thomas J. Wilbanks *(Chair)*, Oak Ridge National Laboratory, Oak Ridge, TN
Barbara Entwisle, Carolina Population Center and Department of Sociology, University of North Carolina, Chapel Hill
Andrew Foster, Department of Economics, Brown University
Roger Kasperson, Department of Geography, Clark University
Edward L. Miles, School of Marine Affairs, University of Washington
Edward A. Parson, University of Michigan Law School, Ann Arbor
Alan Randall, Department of Agricultural, Environmental, and Development Economics, Ohio State University
Peter J. Richerson, Division of Environmental Studies, University of California, Davis
Eugene Rosa, Natural Resource and Environmental Policy, Washington State University
Cynthia Rosenzweig, Climate Impacts Group, NASA Goddard Institute for Space Studies, Brooklyn, NY
Stephen H. Schneider, Department of Biological Sciences, Stanford University
Joel Tarr, Department of History, Carnegie Mellon University
Oran R. Young (ex officio), University of California, Santa Barbara

Paul C. Stern, *Study Director*
Deborah Johnson, *Senior Program Assistant*

Preface

With the growing number, complexity, and importance of environmental problems come demands to include a full range of intellectual disciplines and scholarly traditions to help define and eventually manage such problems more effectively. In the best sense of including talent, insight, and skill from many different places, the National Research Council (NRC) was requested by the U.S. Environmental Protection Agency and the National Science Foundation to help set research priorities for the social and behavioral sciences as these relate to several different kinds of environmental problems. The task was to think broadly and systematically so as to identify a manageable number of promising research questions, the answers to which we believe will contribute to improved environmental decision making. We were specifically cautioned not to promote existing or well-represented and understood research agenda and priorities. Our job was to discover promising new questions and lines of inquiry. Likewise we were asked not to emphasize the field and discipline of economics, mainly on the grounds that it too is well represented and understood in the general environmental realm.

Decision Making for the Environment: Social and Behavioral Science Research Priorities is the result of a 2-year effort by 12 social and behavioral scientists, scholars, and practitioners. Together they represent a wide range of fields and disciplines, including anthropology, decision science, environmental studies, geography, human ecology, management, planning, policy analysis, political science, psychology, resource management, sociology, and urban studies. In addition, the panel benefited from the contributions of dozens of other scientists, scholars, and practitioners from these and other fields, who submitted ideas for our consideration, summarized knowledge in several areas of research, and wrote or reviewed background papers for the study, five of which appear as appendixes to this report.

In making its recommendations the panel met three times in the course of the two-year project, consulted a wide array of professional and disciplinary organizations for suggestions and opinions, and commissioned several papers that pursued in more detail matters of particular interest and relevance to the project.

The panel was ably assisted by the NRC's program officer, Paul Stern, who also participated as a fully vested scholar-member of the panel, based on his long association with the social and behavioral sciences as these relate and apply to environmental issues and problems. Logistical, managerial, and administrative support were well and cheerfully supplied by Deborah Johnson of the NRC's Committee on the Human Dimensions of Global Change.

This report has been reviewed in draft form by individuals chosen for their diverse perspectives and technical expertise, in accordance with procedures approved by the NRC's Report Review Committee. The purpose of this independent review is to provide candid and critical comments that will assist the institution in making its published report as sound as possible and to ensure that the report meets institutional standards for objectivity, evidence, and responsiveness to the study charge. The review comments and draft manuscript remain confidential to protect the integrity of the deliberative process.

We wish to thank the following individuals for their review of this report: William C. Clark, John F. Kennedy School of Government, Harvard University; Howard Kunreuther, Risk Management and Decision Resources Center, Wharton School, University of Pennsylvania; D. Warner North, NorthWorks, Inc., Belmont, California; Ortwin Renn, Chair of Environmental Sociology, University of Stuttgart, Germany; Nigel Roome, Faculty of Social Science, Erasmus University, Rotterdam, The Netherlands; and Elke U. Weber, Graduate School of Business, Columbia University.

Although the reviewers listed above have provided many constructive comments and suggestions, they were not asked to endorse the conclusions or recommendations nor did they see the final draft of the report before its release. The review of this report was overseen by Edward A. Parson, University of Michigan Law School. Appointed by the NRC, he was responsible for making certain that an independent examination of this report was carried out in accordance with institutional procedures and that all review comments were carefully considered. Responsibility for the final content of this report rests entirely with the authoring panel and the institution.

We also wish to thank the following individuals for their review of papers that appear as appendixes of this report: William Freudenburg, University of California, Santa Barbara; Rejean Landry, Laval University; Elke U. Weber, Columbia University; Michael DeKay, Carnegie Mellon University; Asseem Prakash, University of Washington; Clinton Andrews, Rutgers University; Radford Byerly, University of Colorado; Roger Pielke, Jr., University of Colorado; Kathryn Harrison, University of British Columbia; and Richard Morgenstern, Resources for the Future, Washington, DC.

The panel is well aware of the limitations of so-called priority-setting exercises, such as the one we just completed. There is no way to be totally comprehensive, for there are simply too many combinations of fields, disciplines, specialties, and problem types for any small group such as this panel to cover thoroughly. Nonetheless, we made strong efforts to reach out and to consult with a wide and diverse a group of professionals and organizations. There is no way to identify and then promote all worthy research possibilities. Indeed, the sponsoring agents for this project were very clear in their charge to consider and focus on a manageable few topics—but to justify these choices in a clear and rational manner so that anyone would be able to appreciate why we decided as we did. Here we have been most diligent, as I trust the reader will agree very soon after beginning to read our report.

If some of our recommendations gain the attention we believe they warrant and in time secure sufficient research support to answer the questions posed, then our collective efforts on this panel will have been well worth the time, effort, and energy put into this project.

Garry D. Brewer, *Chair*
Panel on Social and Behavioral Science Research Priorities for Environmental Decision Making

Contents

Executive Summary

The social and behavioral sciences provide an essential but often unappreciated knowledge base for wise choices affecting environmental quality. These sciences can help decision makers of all kinds to understand the environmental consequences of their choices and the human consequences of environmental processes and policies, as well as to organize decision-making processes to be well informed and democratic.

Recognizing the need to develop more fully the social and behavioral science knowledge base for environmental decision making, the U.S. Environmental Protection Agency (EPA) and the National Science Foundation (NSF) asked the National Academies to identify a few science priorities—areas in which concentrated new research efforts could both advance the environmental social and behavioral sciences and contribute to improved decisions affecting environmental quality. The National Academies were asked to focus primarily on the social and behavioral sciences other than economics, because they have not received much attention from environmental decision-making organizations, and to recommend research areas that scored well on three criteria: the likelihood of achieving significant scientific advances, the potential value of the expected knowledge for improving decisions that have important environmental implications, and the likelihood that the research would be used to improve those decisions. We were also asked to consider recommending ways to overcome barriers to the use of research that would have high priority if such barriers could be overcome and invited to make general recommendations for infrastructure that could increase the likelihood that the recommended knowledge across several fields will be used. This report is addressed to two main audiences: potential researchers and potential sponsors of research.

We contacted many research communities in our search for research areas to recommend and considered each suggestion in light of the above criteria. In order to consider the likelihood that research results would be used, we also reviewed available research on the use of scientific results by various kinds of decision makers. We recommend five science priorities that strongly meet the decision criteria.

IMPROVING ENVIRONMENTAL DECISION PROCESSES

Federal agencies should support a program of research in the decision sciences addressed to improving the analytical tools and deliberative processes necessary for good environmental decision making. Decisions affecting environmental processes are among the most challenging facing humanity because of the conjunction of several decision attributes, such as complexity, uncertain and conflicting values, incomplete and uncertain knowledge, long time horizons, high stakes, multiscale management, linkages among decisions, and time pressure. Good environmental decision making requires not only good environmental science, but also improved understanding of human-environment interactions and development and implementation of decision-making processes that integrate scientific understanding with deliberative processes to ensure that the science is judged to be decision relevant and credible by the range of parties interested in or affected by the decisions. The recommended research would use decision science methods to enable environmental decision processes to become increasingly responsible, competent, and socially acceptable. It would build on a foundation of basic research on decision processes, which we assume will continue to receive support. The effort would have three components.

Developing criteria of decision quality. We recommend research to define decision quality for practical environmental decisions. It would consider such questions as: Which characteristics of decision processes are associated with judgments of decision quality or acceptability by decision participants and observers? Do different kinds of people apply different criteria of decision quality? To what extent does increased attention to ideals of good public decision processes yield more positive assessments of actual decision quality? Are decisions of higher normative quality associated with preferred social and environmental outcomes? How can research results on such questions best be disseminated to their potential users?

Developing and testing formal tools for structuring decision processes. We recommend research to refine and apply tools from the decision sciences for helping decision makers better approximate ideals of good decision processes. The research might address such questions as: How can formal methods of value elicitation be applied effectively in real world decision settings? How can judgments about the nature and likelihood of a range of outcomes be made more routine and workable through the use of information technologies? What systematic methods for arriving at collective preference can be applied in realistic environmental decision settings that can complement those of social benefit-cost analysis and that do not adopt problematic assumptions typical of that approach? How can learning be built into decision procedures to allow for updating over time? How can risk communication methods be used to make issues of preference and uncertainty intelligible and useful to key decision makers and affected parties? How can decision-aiding approaches help individuals by structuring the values, uncertainties, and broader implications of their choices?

Creating effective analytic-deliberative processes. We recommend research to strengthen the scientific base for organizing processes, such as are now being used with increasing frequency in government, in which a broad range of participants take important roles in environmental decisions, including framing and interpreting scientific analyses. The recommended research would address such questions as: What are good indicators for key attributes of success for analytic-deliberative processes, such as decision quality, legitimacy, and improved capacity for future decision making? How are these outcomes affected by the ways the processes are organized, the ways they incorporate technical information, and the environmental, social, organizational, and legal contexts of the decision at hand? How can decision processes be organized to ensure that all sources of relevant information, including the local knowledge claims of nonscientists, are gathered and appropriately considered? How can these processes be organized to reach closure, given the challenges of diverse participants and perspectives? How can decision-analytic techniques be used to best advantage in these decision processes? How can technical analyses be made transparent to decision participants who lack technical training?

The recommended research would advance understanding of the characteristics of good decisions, further develop decision science tools for practical uses, and advance theoretical and practical understanding of ways to inform decisions through analytic deliberation. It would offer scientific guidance to the growing numbers of federal agencies and others who are opening environmental decision making to a range of stakeholders and affected parties as to how best to make these processes serve societal goals.

INSTITUTIONS FOR ENVIRONMENTAL GOVERNANCE

Federal scientific and environmental agencies should support a concerted effort to build scientific understanding needed for designing and evaluating institutions for governing human activities that affect environ-

mental resources. The question of environmental governance is typically posed as a choice among a few basic policy strategies, such as direct management or regulation by centralized government agencies, market-based governance that relies on privatizing certain rights and allowing markets to emerge for them, and strategies that devolve authority to voluntary action or to organizations at state or local levels. Research indicates that no governance structure works best for all situations; rather the critical task is to find the arrangement that is most appropriate for particular governance problems. The recommended research would link the research traditions of policy analysis and evaluation with a research tradition that analyzes environmental policies in terms of institutional design. It would address ways of meeting the key requirements for adaptive governance of complex systems of human-environment relationships, such as providing information, managing conflict, inducing rule compliance, providing physical and informational infrastructure, and providing flexibility to adapt to change. It would also address ways to design context-specific property rules, build legitimacy and trust when facing complex and large-scale environmental problems and multiple interested publics, and develop institutional forms that cross scales of organization.

This research priority, which has been identified in several previous National Research Council reports, would bring together two separate research traditions on environmental governance. It would elaborate a conceptual framework and supporting bodies of knowledge that environmental policy makers, natural resource managers, and other participants in environmental governance could use to improve resource management institutions and to design more effective linkages among institutions at different levels of governance.

THE ENVIRONMENT IN BUSINESS DECISION MAKING

Federal agencies should substantially expand support for research to understand the influence of environmental considerations in business decisions. Although business decisions are among the dominant forces in humanity's impact on the environment, and although many of these decisions create societal commitments that are difficult if not impossible to reverse, the role of environmental considerations in business decision making has been surprisingly and seriously understudied. Several research directions are highly promising.

Environmental performance and competitive advantage. When does it pay for businesses to be "green"? When it does pay, to what extent does competitive advantage come from external incentives or from characteristics of firms? Do the pro-environmental practices of leading firms diffuse to other businesses, or do they just segment the market and have little broader effect? Why do some firms fail to adopt pro-environmental practices that would offer attractive rates of return?

Effects of demand on environmental performance. Under what conditions is the environmental performance of firms driven by demands from current customers? Current or potential investors? International or emerging markets? Influential business partners?

Effects of supply chains and production networks. Under what conditions do the demands of dominant businesses affect performance throughout their supply chains? How can supply chain mandates leverage environmental performance? How can product life-cycle analysis aid environmental decision making at the level of supply chains?

Sectoral standard-setting. How can trade associations effectively regulate the environmental decision making of their members? Which industries are amenable to this kind of influence to greater or lesser extent, and why?

Decision factors in industrial ecology. Under what conditions have industrial innovations reshaped entire materials chains to reduce extraction and waste production? How have decisions about single technologies influenced entire systems of energy and materials transformation and created opportunities or barriers to the development of more closed-cycle industrial systems?

Environmental accounting and disclosure practices. How can the effects of environmental performance on economic performance be measured more effectively? How can environmental disclosure practices be standardized to enable better accounting?

Government policy influences on business decision making. How can environmental policies designed to create incentives for green innovation avoid privileging particular technologies? How can voluntary initiatives be combined effectively with other policy instruments?

This science priority would begin to integrate several bodies of research being pursued by a growing number of researchers to generate knowledge of value to policy makers in government and the private sector who want to improve the environmental performance of businesses.

ENVIRONMENTALLY SIGNIFICANT INDIVIDUAL BEHAVIOR

Federal agencies should support a concerted research effort to better understand and inform environmentally significant decisions by individuals. Because the activities of individuals and households have major environmental consequences in the aggregate, considerable environmental improvement can in principle result from change in their behavior. However, fundamental understanding is only beginning to develop regarding how various influences interact to shape and alter that behavior, and we lack good measures of environmentally significant consumption. Research in four specific areas could provide usable results in the relatively near term.

Indicators of environmentally significant consumption. The recommended research would link measures of environmental impact, such as of energy and materials transformations and life-cycle impact of products and activities, to important individual choices. Careful accounting studies that combine physical science expertise and knowledge of human behavior can provide individuals with better understanding of which choices really make an environmental difference.

Information transmission systems. Recommended research would address ways that information transmission systems, including networks of information sources and ways of producing and validating indicators, affect the likelihood that audiences will receive accurate information about the environmental implications of their choices from trusted sources when they need it.

Integration of information with other policy instruments. The effects of information on behavior depend on incentives and infrastructure, and vice versa. The recommended research concerns the joint effects of information and other policy instruments, in particular individual-behavior contexts, such as transportation mode choice, investment in energy efficiency, and management of household wastes.

Fundamental understanding of consumer choice and constraint. Recommended research would build a basis of fundamental knowledge of the ways in which personal factors (values, attitudes, skills, etc.) and contextual factors (economic costs, properties of the built environment, government policies, etc.) combine to influence various types of environmentally significant consumer choices.

The recommended research will inform decision makers at various levels who want to understand and anticipate changes in environmentally significant individual behavior or to use information and other policy tools to promote socially desired environmentally significant behavior. It is also likely to lead to practical understanding relevant to other areas of policy and to better fundamental understanding of individual behavior under complex real-world conditions and of the determinants of environmental resource consumption.

DECISION-RELEVANT SCIENCE FOR EVIDENCE-BASED ENVIRONMENTAL POLICY

To strengthen the scientific infrastructure for evidence-based environmental policy, the federal government should pursue a research strategy that emphasizes decision relevance. The strategy should encompass four substantive research elements: (1) developing indicators for environmental quality, including pressures on the environment, environmental states, and human responses and consequences, that are designed to serve the needs of decision makers; (2) making concerted efforts to evaluate environmental policies; (3) developing better methods for identifying the trends that will determine environmental quality in the future; and (4) improving methods for determining the distributional impacts of environmental policies and programs.

Major research efforts in environmental science are often justified by their societal relevance. Such efforts typically produce high-quality science, but they have repeatedly fallen short in addressing the questions most important to societal decision makers. This science priority would enable federal agencies to greatly improve the infrastructure of scientific information and methods toward the goal of informing practical decisions. The recommended scientific activities would integrate the social sciences and the natural sciences of the environment and would address both environmental conditions and their human connections. They would help inform practical decision making while also informing scientific research and help increase the influence of science in environmental decisions relative to the influences of politics and ideology.

Processes for determining which research is most decision relevant should be participatory: choices about how to construct indicators, evaluate policies, and so forth should be made with the participation of the full range of likely users of measures, evaluations, and analyses. These choices are not purely technical. Measurement embodies values about what is most worthy of attention, a matter on which affected parties often disagree. Choices about what evidence to collect for policy are probably most appropriately made through broad-based analytic-deliberative processes.

Federal agencies should work to make environmental science more decision relevant in each important area of environmental policy. The effort should involve the following activities:

Improving human-environment indicators. Indicators are essential for making sense of an overwhelming amount of environmental information, but no set of environmental indicators in the United States commands the respect and attention of the public or policy makers. An integrated effort based on the following principles can change this situation.

- Social science and natural science research should be integrated in a comprehensive approach to developing indicators that are relevant and usable for environmental policy. These indicators should cover not only states of the biophysical environment, but also human influences on nature (pressures on the environment, such as population, technology, consumption, and pollutant emissions) and the impact of the physical world on humans, including public and private actions taken to reduce pressures, protect states, and adapt to environmental changes, as well as the human consequences of environmental events, taking responses into account.

- To ensure the decision relevance and comprehensibility of indicators, government agencies involved in developing them should create them in collaboration with the producers and potential users of the information, including a variety of nonscientists.
- Good indicators require close collaboration among existing organizations and may require the creation of new ones. Recommendations to create a federal Bureau of Environmental Statistics deserve serious attention because of this need for collaboration.
- Special efforts may be required to enable rapid development of useful indicators under conditions of surprise or disaster, when existing indicators are inadequate.

Efforts to develop indicators, regardless of the environmental system or problem that requires measurement, should entail the following steps: (1) identifying the user audience and the uses to which the indicators will be put; (2) assessing and evaluating existing efforts and indicators; (3) developing new methods for indicator construction, if necessary; (4) identifying the data needed for the indicators and evaluating their availability; (5) pilot testing each indicator to analyze how well it meets the specified uses.

Environmental policy evaluation. Federal agencies should support a concerted research effort to evaluate the effectiveness of environmental policies established by public and private actors at the international, national, state, and local levels. This research would apply techniques of evaluation research that have been used primarily to assess the effectiveness of social welfare policies to the domain of environmental protection. It would examine the outcomes of environmental regulations and other environmental policies in terms of effectiveness, efficiency, fairness, and public acceptability and strengthen methods and capacity for determining the results of environmental policies.

Improving environmental forecasting. Federal environmental agencies should undertake an assortment of research initiatives to collect, appraise, develop, and extend analytic activities related to forecasting in order to improve environmental understanding and decision making. As with the development of indicators, forecasting efforts should focus from the start on the human setting of environmental decision making, should encompass human influences on the environment as well as biophysical processes, and should be directed at decision-relevant outcomes, including environmental, health, and socioeconomic outcomes and the distribution of these outcomes across segments of the population. We specifically recommend support for efforts to identify best practices in forecasting, for continuing environmental modeling forums patterned on the Energy Modeling Forum at Stanford University, and for improving ways to describe uncertainties in forecasts.

Determining distributional impacts. Federal agencies should support concerted efforts to improve the data, methods, and analytic techniques for determining the distributional impacts of environmental policies and programs related to issues of environmental inequities and their abatement. These efforts should include research to determine the most appropriate levels of social, spatial, and temporal aggregation of measurement for environmental monitoring and indicator development and should address the following themes: defining key variables (e.g., minority population), analyzing dependence of impacts on spatial and temporal scale; developing integrated biophysical and social models that include multiple stressors, multiple exposure pathways, and social vulnerability; and improving visualization and risk communication regarding the impacts of environmental conditions and policies.

Research to develop the scientific foundation for evidence-based environmental policy would enable major advances in fundamental understanding of the dynamics of human-environment interaction by vastly increasing the possibility of analyzing these relationships quantitatively. It would also greatly increase the decision relevance of environmental analyses by providing credible measures and methods of analysis for addressing issues of critical concern to both decision makers and scientists.

1

Introduction

The natural sciences provide an essential knowledge base for wise choices when human activities may have significant environmental consequences. Only through these sciences can decision makers understand the dynamics of environmental systems and the ways in which human actions reverberate through these systems to affect environmental quality. The social and behavioral sciences also provide an essential knowledge base, although their roles are not as commonly recognized or as fully institutionalized in environmental policy. Only through these sciences can decision makers understand which policies will induce the desired human actions in relation to the environment and what ultimate effects environmentally important decisions are likely to have on human well-being. Moreover, environmentally important decisions may themselves be improved with better application of behavioral and social scientific knowledge.

Recognizing the need to develop these kinds of knowledge, the U.S. Environmental Protection Agency (EPA) and the National Science Foundation (NSF) asked the National Academies to identify a few science priorities that could contribute to improved environmental decision making and also advance the social and behavioral sciences. The National Academies were given a broad purview for this study. The study was to consider all relevant social and behavioral scientific perspectives and approaches, both on their own and as they relate to and integrate with perspectives from the natural sciences, engineering, and mathematical sciences. It was to consider research on decision processes in government organizations and elsewhere, as well as research on human-environment relationships that might have practical value for decision making. It was to consider areas of research that could improve environmentally important decisions regardless of whether the decision makers are government agencies, private companies, other organizations, or individuals. It was to focus, however, on the social and behavioral sciences other than economics because they have not received much attention from decision-making organizations, and to define research areas that would build on strengths in these sciences and link them with each other, with economics, and with the natural sciences so as to produce a deeper understanding of environmental issues. We understood the relevant social and behavioral sciences to include the traditional disciplines of anthropology, geography, political science, psychology, and sociology as well as various associated interdisciplinary fields, such as decision science, communications research, policy sciences, human ecology, and science and technology studies. Thus, we did not consider recommending priority research areas that we judged to fall primarily in economics, regardless of how well those areas might score against the decision criteria we used. This report has two main audiences: potential researchers and potential research sponsors.

The National Academies were asked to recommend research areas that score well as measured against three criteria: the likelihood of achieving significant scientific advances, the potential value of the expected knowledge for improving decisions having important environmental implications, and the likelihood that the research would be used to improve those decisions. They were also asked to consider recommending ways to overcome barriers to the use of research that would have high priority if such barriers could be overcome and invited to make general recommendations for infrastructure that could increase the likelihood that the recommended knowledge across several fields will be used.

HOW WE CONDUCTED THE STUDY

The National Academies organized the study under the auspices of its Committee on the Human Dimensions of Global Change, which has since 1989 advised federal agencies on research issues in the area of human-environment interactions and has produced several previous reports identifying promising research directions (National Research Council, 1992, 1994b, 1999b). The committee participated in selecting the membership of the panel and in reviewing this report. Panel members were selected to include expertise from across the social and behavioral sciences, with strong representation of researchers grounded in these disciplines who are engaged in studying environmental issues. Members also include individuals with backgrounds in the environmental natural sciences and engineering, experience in governmental and private organizations whose decisions have significant environmental impact, and expertise in the use of science in policy and organizational decision making.

Identifying Science Priorities

From the outset, we decided to look widely for ideas about research areas that might meet the three decision criteria. Before our first meeting, we sent a message to e-mail lists of all the relevant research groups and networks that we could identify in which we explained the panel's task, listed our decision criteria, solicited suggestions of research areas from the recipients, and invited them to pass the request along to anyone they thought might have worthy ideas for us to consider.[1] We considered all the suggestions that were submitted, along with suggestions from panel members and sponsors at our initial meeting in June 2003, and identified about a dozen broad research fields in which priority areas might lie. We then invited a scientist working in each broad field to write a short memorandum identifying research areas in that field that he or she thought met our decision criteria and to explain this judgment.[2] We invited these individuals to discuss their memoranda with us at our second meeting. After that meeting, we invited some of them to expand these into papers that appear as appendices to this report. At our third meeting, we refined our focus to the five recommended research priorities described in Chapters 2 through 6.

Applying the Decision Criteria

The three decision criteria that the panel was given entailed making predictive judgments about the consequences for science and decision making of differential investments in research. Historical examples of important advances in the social and behavioral sciences suggest the difficulty of predicting the path, impacts, costs, and benefits from innovations in the social sciences. If would have been exceedingly difficult, if not impossible, to predict how Garrett Hardin's famous paper on the "tragedy of the commons" (Hardin, 1968) would have affected researchers in many different fields or stimulated the search for management regimes that do not result in degradation of common property. Although we still do not have definitive answers in this search, much has been learned about what kinds of arrangements tend to work in certain contexts and why they work (e.g., National Research Council, 2002a; Dietz, Ostrom, and Stern, 2003). Certainly the full body of research on problems of managing the commons has had a notable impact, but it would have been very difficult if not impossible to anticipate the nature of the impact or who would use the knowledge. Similarly, it would have been difficult to predict how the development of the field of applied welfare economics from the 1920s to the 1950s, including the evolution of its many analytic tools in the 1960s and 1970s that together comprise social benefit-cost analysis, would affect environmental decision making today.

These examples suggest that despite the admirable logic of the decision criteria, they are better suited to considering information that is provided for a narrowly defined application than for assessing the potential of basic research or research with a broad range of potential intellectual implications and practical applications. The panel sought nevertheless to discipline the study by repeated reference to the decision criteria, to develop collective judgments in relation to the criteria as conscientiously and consistently as possible, and to seek guidance from past experience and empirical research on the use of knowledge from the social and behavioral sciences in environmental and other practical decision making.

We consulted panel members with expertise in decision processes for advice on what procedure to use to judge potential science priority areas against the criteria. With their advice, we decided that the criteria as given to us were sufficiently vague (and the topics sufficiently unformed) as to make it highly likely that if we used a procedure of voting or subjective weighting of topics against criteria, the results might not be meaningful because different panel members would have interpreted the criteria differently. Consequently, we decided to specify the criteria further by identifying a number of factors that are likely to act as means to the ends outlined in the criteria (described below). Each panel member agreed to consider how each of these contributing factors applied to each research area and to judge that area accordingly against the relevant decision criterion. At the third meeting. we engaged in a discussion of each of the previously identified topics (now accompanied by draft text that ensured a common understanding of what the topic covered) in light of the criteria and the contributing factors in the hope of reaching consensus on which topics deserved inclusion among the science priorities and which, for whatever reasons, did not. The chair held in reserve the option of using a subjective weighting scheme if discussion failed to reach consensus. That option turned out to be unnecessary, as we readily reached consensus. Once it was agreed which topics deserved inclusion, we worked to frame a set of no more than five science priorities that would coherently include the topics that met the criteria. The overall process of selecting topics and applying the decision criteria involved winnowing, combining, and reformulating the topics that had initially been proposed to arrive at the final list.

Likelihood of Achieving Significant Scientific Advances

We rated potential science priorities highly on this criterion when we judged the following factors to be applicable:

- The research community is ready and able to conduct the research (e.g., concepts, methods, and data are available but not yet adequately applied in this area).
- Successful research would provide new frameworks for thinking or sources of understanding (e.g., data, methods) that could lead to advances in environmental decision making over time.
- Successful research would overcome or reduce gaps in knowledge or skill that now inhibit opportunities for improved environmental decisions in a given context.

Potential Value of the Expected Knowledge

We rated potential science priorities highly on this criterion when we judged the following factors to be applicable:

- The research findings are relevant for decisions with important environmental consequences and social or economic implications that are significant to affected parties or governments.
- The research findings are relevant for a diverse range of environmental decisions.
- The research results have significant potential to create, compare, and implement more attractive policy alternatives.

Likelihood That the Research Would Be Used

We informed our judgments of proposed science priorities in relation to this criterion by examining empirical research on the use of scientific findings by various kinds of decision makers. Much of this research focuses on the use of information from the social sciences in government and private-sector organizations (see Appendix A for a review and an annotated bibliography). We also considered the results of research on the factors affecting individuals' use of information in making environmentally relevant decisions (for reviews, see Gardner and Stern, 2002; National Research Council, 2002b; see also Chapter 5).

The research suggests that the likelihood of use of social science research is affected by attributes of the decision-making organizations, the researchers' activities, and the links between researchers and users.[3] Studies suggest that decision-making organizations are more likely to use science that they have expressly requested, particularly from internal sources; when unfamiliar problems arise; when there are incentives for seeking information; and when they believe the research can provide authoritative support for their decisions (e.g., Oh, 1996b; Oh and Rich, 1996). Information users' acquisition efforts are also important in getting research used (Landry, Lamari, and Amara, 2003). The pattern of scientific challenge, in which research results are more likely to be challenged if they threaten well-organized interests, affects the likelihood that scientific findings will be accepted by public policy makers (Freudenburg and Gramling, 2002). Characteristics of the research itself, including technical quality and the extent to which the research directly focuses on user needs rather than on advancing scientific knowledge, are not consistently related to utilization (Landry et al., 2003). However, researchers' efforts to disseminate results and to adapt their reporting to users' needs are strongly associated with increased use of the information in some studies (Greenberg and Mandell, 1991; Landry, Amara, and Lamari, 2001; Landry et al., 2003). In addition, research use is facilitated by formal or informal links between researchers and research users (Huberman, 1990; Landry et al., 2001, 2003). Because of the small number of empirical studies on these issues, however, the generality of the findings is uncertain.

We rated potential science priorities highly on the criterion of likely use of research results if we judged the following factors to apply:

- Decision makers, such as those in organizations that make environmentally important decisions or among groups affected by such decisions, would be likely to request the research or the information it can yield.
- Decision makers, including parties affected by decisions, have incentives to seek and use the information, for example, to help them achieve personal, group, or organizational objectives.
- Researchers have incentives to disseminate their findings in ways that usually reach potential users and not only to academic publication outlets.
- Good organizational links or intermediaries exist that provide lines of communication or "translation" services between the likely producers and the likely users of the research results.

The above factors favoring the use of research results are not external to the decision-making process. Organizations that support or use research can act to create favorable conditions for using research when those conditions do not already exist. In some of the recommended science priority areas, we have recommended such actions.

THE RESULTS

Science Priorities

The panel recommends five science priorities for improved environmental decision making that, in our judgment, strongly meet the selection criteria we have been given. They are described in detail in Chapters 2 through 6, along with explanations of how the recommended research can improve decision making. We note the priorities briefly here.

Environmental decision processes. We recommend a program of research in the decision sciences addressed to improving the analytical tools and analytic-deliberative processes necessary for good environmental decision making. It would include three components: developing criteria of decision quality; developing and testing formal tools for structuring decision processes; and creating effective processes, often termed analytic-deliberative, in which a broad range of participants take important roles in environmental decisions, including framing and interpreting scientific analyses.

Institutions for environmental governance. We recommend a concerted effort to build scientific understanding needed for designing and evaluating institutions for governing human activities that affect environmental resources. This science priority, which has been identified in several previous National Research Council reports and by the National Science Foundation, would bring together the research traditions of policy analysis and institutional analysis to elaborate science-based tools that participants in environmental decisions can use to improve resource management institutions and to design more effective linkages among institutions at different levels of governance.

Green business decision making. We recommend substantially expanded support for research to understand the influence of environmental considerations in business decisions. The research would address such issues as when and under what conditions better environmental performance provides competitive advantages; how the demands of customers, suppliers, and investors affect environmental performance; how environmental outcomes may be affected by changes in business supply chains; and how environmental accounting procedures and sectoral standard-setting activities can affect environmental outcomes.

Environmentally significant individual behavior. We recommend a concerted research effort to better understand and inform environmentally significant decisions by individuals. This priority includes research in four specific areas: indicators of environmentally significant consumption, fundamental research on consumer choice and constraint, transmission systems for decision-relevant information for individuals, and integration of information with other policy instruments.

Decision-relevant science for evidence-based environmental policy. We recommend that the federal government strengthen the scientific infrastructure for evidence-based environmental policy by pursuing a research strategy that emphasizes decision relevance. It should do this by developing decision-relevant indicators for environmental policy, including pressures on the environment, environmental states, and human responses and consequences; by making concerted efforts to evaluate environmental policies; by developing better methods for identifying the trends that will determine environmental quality in the future; and by improving methods for determining the distributional impacts of environmental policies and programs. These efforts will require integrating the social sciences and the natural sciences of the environment. Decisions about how to construct indicators, evaluate policies, and so forth will require the involvement of the full range of parties affected by environmental decisions because these choices are not purely technical. Measurement focuses attention on what has been measured, and affected parties often disagree about what is most worth measuring, which outcomes of policies are most important, and the like.

Cross-Cutting Issues

Because we were asked to identify a very small number of science priorities, we have not mentioned many other intriguing and meritorious topics. Here we highlight three topics that, although we have not identified them as separate science priorities, are so pervasive and so linked to several of the science priorities that they warrant attention across the science priority areas.

Innovation and Technological Change

Research on innovation has a long tradition and many different disciplinary sources, including psychology, history, anthropology, political science, economics, geography, and sociology. Many core concepts are shared broadly. For example, the ideas of "first movers" and "late adopters" one encounters in a corporate strategy or management journal have equivalents in each of the other disciplines just noted. Concepts of evolution and adaptation in innovations over time are also widespread, although interdisciplinary awareness of them is not (Erwin and Krakauer, 2004). Early adopters typically pay a premium in economic and other terms compared with those who come later. Innovations may work in some cultures but not in others for any number of empirical, researchable reasons. Issues of communication and education also appear routinely in studies of innovation across disciplines and fields of human endeavor. Innovation research offers a potentially fruitful approach for understanding and improving environmental decision making.

Innovation is important for environmental policy both because of the role of technological innovation in creating and ameliorating environmental problems and because of the need for policy innovation at all levels and for its diffusion. Issues of innovation are particularly important to our science priorities in the areas of decision-making processes, environmental governance, green business decision making, and individual behavior. Concepts from research on innovation and technological change can usefully be applied in all these priority areas.

System Complexity

Human-environment systems and the policy systems used to govern them are both highly complex. Researchers who study complex systems have developed a variety of concepts that can be useful for understanding these systems and improving their functioning. Consider, for example, the ways the highly capitalized and complex transportation systems upon which modern societies depend may resist the transformations required to sustain these societies into the future. The problem has been termed "technological lock-in" by systems theorists and is reflected in ongoing discussions about the future of a hydrogen-based economy. One can readily build a hydrogen-powered car today, but the technology system required to get the hydrogen to it on a mass market basis does not exist. The gasoline engine is "locked in" by the fuel manufacture and distribution infrastructure. Change is still possible, but it will take much longer, is more complicated, and runs a great risk of generating unanticipated consequences as it ripples through the coupled technological, economic, and social systems (Bijker, Hughes, and Pinch, 1987).

For comparatively less highly capitalized and simpler systems, in which enabling technologies are not tightly coupled, system changes may be easier to come by (Shapiro and Varian, 1999). For example, one can usually change air scrubber technologies without affecting any of the underlying manufacturing technologies, since these are only loosely coupled. This is not the case with a core manufacturing technology such as the use of lead solder in electronics manufacturing, which is tightly coupled to other technologies (Allenby, 1992).

The resilience or brittleness of systems matters too. Resilient systems are capable of absorbing or otherwise dealing with external threats and opportunities. Brittle ones are usually less capable. Proposals to shift to "distributed generation" of energy and electricity arguably underestimate the brittleness of existing and facilitating infrastructures in the face of change. Brittleness need not preclude innovation, however, as the story of the cell phone and its winning battle with conventional telecommunications indicates so well.

These observations suggest that research that provides environmental decision makers with better understanding of complex systems and their evolution may have widespread value. Organizational studies with a complex systems perspective, especially multisectoral ones exploring the roles and relationships of private, public, not-for-profit, and nongovernmental institutional forms as these relate to environmental innovation and technological change, are likewise attractive. Thus, a complex systems perspective may be usefully applied in the recommended priority areas of improving decision processes, environmental governance, and business decision making.

Combining Social and Natural Science

Even though our task was to identify priorities that flowed out of the social and behavioral sciences, each of our recommended science priorities requires collaboration, and sometimes integration, across the social and natural sciences. Each one builds on measurement and analysis of both biophysical and human conditions and processes, as well as of human-environment interactions. Coupling the social and natural sciences is an increasingly important element of emerging research and development programs in the federal agencies. For example, the NSF's new cross-directorate program on environmental research and education (ERE) explicitly emphasizes the coupling of human and natural systems and of people and technology (Pfirman and the Advisory Committee for Environmental Research and Education, 2003). It emphasizes as principal research questions "how the environment functions, how people use the environment, how this use changes the environment, . . . and how the resultant environmental changes affect people" (p. 13). Efforts to implement our recommendations would therefore contribute to efforts at NSF, in federal environmental agencies, and elsewhere to develop the multidisciplinary science needed to inform environmental decisions.

Considering the parallels between this study and forward-looking research planning efforts in federal agencies, it is not surprising to see numerous substantive overlaps in recommendations. For example, the NSF-ERE report identifies numerous recommended research areas, including (Pfirman and the Advisory Committee for Environmental Research and Education, 2003):

- Identifying decision processes that effectively combine analytical, deliberative, and participatory approaches to environmental choices, which will guide scientists and engineers toward the generation of decision-relevant information (p. 35).

- Understanding the patterns and driving forces of human consumption of resources, and identifying policies and practices that influence materials and energy use decisions, including incentives (p. 33).
- Conceptualizing and assessing the role that institutions play in the use and management of global, national, and local common-pool resources and their associated environmental conditions (p. 36).
- Developing decision-making strategies and institutional approaches to most effectively solve problems and deal with uncertainty (p. 35).

Close parallels to each of these recommendations and others from the ERE advisory group can be found in the present study. Our recommendations in turn are completely consistent with the recognition of the NSF-ERE group and others (e.g., National Research Council, 2001a) that an adequate decision-relevant understanding of the environment depends on better coordination across the sciences.

NOTES

1. We sent this request to the following groups, associations, and networks that include social and behavioral scientists with environmental interests: the Society for the Psychological Study of Social Issues and the Division of Population and Environmental Psychology of the American Psychological Association; the technology and environment politics section of the American Political Science Association; the environment and technology section of the American Sociological Association, the Rural Sociology Society, the Society for the Study of Social Problems, the environmental studies section of the International Studies Association, the Association for Public Policy Analysis and Management; the International Society for Ecological Economics; the social, economic, and political sciences and societal impacts of science and engineering sections of the American Association for the Advancement of Science; the International Association for the Study of Common Property; the Society for Human Ecology; the International Association for Society & Natural Resources; the risk communication section of the Society for Risk Analysis; the International Society for Industrial Ecology, and the Organizations and the National Environment network. Individual panel members also sent the request to listserv groups they knew of with similar interests.

2. The list of fields, as framed after the first meeting, was: individuals and complex information; business decisions and the environment; barriers to use of (social) science by policy makers; forecasting and forecasting tools; technological change; institutions for environmental management/governance; evaluation of environmental policies and activities; environmental indicators; vulnerability and the distribution of risks; improving environmental decision processes; and environmental governance outside governments. We invited the experts to interpret these brief descriptions in ways that would allow them to present the research areas they considered most promising according to our criteria. We continued to refine the list of fields as the study proceeded.

3. The research we reviewed on the likelihood of use of research consisted mainly of studies focused rather narrowly on the links between particular scientific products and that inform relatively specific decisions. There is another tradition of social scientific studies of

science that examines more broadly the role of science and scientists in a variety of decision and policy processes (e.g., Brunner and Ascher, 1992; Gunderson, Holling, and Light, 1995; Sarewitz, Pielke, and Byerly, 2000; van Asselt, 2000; Freudenburg and Gramling, 2002; Ascher, 2004). Although this tradition contains useful insights regarding how science is used and misused in policy processes, we did not find it useful for judging which research areas are most likely to produce knowledge that will be used.

2

Improving Environmental Decision Processes

Federal agencies should support a program of research in the decision sciences addressed to improving the analytical tools and deliberative processes necessary for good environmental decision making. This research effort would have three components: (a) developing useful criteria to characterize and evaluate the quality of environmental decisions, (b) developing and testing formal decision science tools for structuring decision processes, and (c) building and testing concepts and practice for broadly based analytic-deliberative processes. Basic research on decision processes in individuals, groups, and organizations provides an essential foundation for this science priority. The National Science Foundation has supported such research in the past, sometimes in conjunction with the U.S. Environmental Protection Agency (EPA), and we expect support for basic decision research to continue. Our emphasis here is on research that would employ and advance basic understanding for the practical objective of improved environmental decision making.

THE RESEARCH NEED

Individuals, organizations, and ultimately societies, through their choices, have significant effects on the natural environment. The large human footprint on the Earth and the potential for huge mistakes make it imperative that the major choices are well informed and adequately considered. Decisions affecting environmental processes, however, are among the most challenging facing humanity because of the following collection of attributes that environmental choices usually share:

- *Structural complexity*: choices affect phenomena that operate at multiple scales; decision-making entities also exist at multiple scales, not necessarily matched to those of the phenomena; and many different kinds of expertise are required to understand the issues.
- *Multiple, conflicting, and uncertain values*: people affected by the choices have deeply held values often tied to spiritual, cultural, stewardship, or equity concerns that they are unwilling to negotiate or trade off; people differ in their value priorities; and sometimes their values seem to shift unexpectedly.
- *Long time horizons*: the consequences of choices made now may extend for decades or longer.
- *Open-access structure*: it is often difficult to exclude people from using or polluting a resource, putting that resource at considerable risk of overuse and decline (see Chapter 3).
- *Incomplete and uncertain knowledge*: the consequences of choice options may be unknown or in dispute among scientists; they may also be dependent on ongoing processes of social or environmental changes that are also little understood.
- *High stakes*: the long-term implications of the wrong choice for environment and society may be profound.
- *Time pressure*: decisions must be made without waiting for scientific certainty or agreement on values.

These points are well recognized by observers of environmental decision processes (e.g., Funtowicz and Ravetz, 1992; National Research Council, 1996; Dietz and Stern, 1998; Renn, 2003). A further challenge is to address the linked nature of environmental processes and environmental decisions across time scales, physical scales, and institutional scales. Decisions made at one scale can be transformed or undermined by processes at other scales, which must therefore be taken into account. Researchers have only recently given serious consideration to this challenge to environmental decision making and management (Cash and Moser, 2000; Young, 2002; Berkes, 2002; also see Chapter 3).

In addition, environmental choices are affected by decision makers' attention to various environmental or other aspects of the choices. Individuals' apparent preferences shift depending on how choices are framed and on their interpretations of and affective reactions to information (Tversky and Kahneman, 1981; Kahneman, Ritov, Jacowitz, and Grant, 1993; Slovic, 1995), and the apparent priorities of organizations and governments shift as a function of how interested parties shape decision agendas (Kingdon, 1987).

Decisions of such difficulty require a variety of inputs. Elected representatives, who are normally entrusted with making value choices, rarely have sufficient expertise to make well-informed decisions, but scientific and technical experts are not well suited or trusted to address the value issues. To meet the challenges, scientists have developed numerous analytical tools to inform decision makers about the functioning of environmental systems and the likely consequences of available choices. Mathematical models of complex systems represent the multiple layers and linkages that constitute environmental systems and forecast the consequences of interventions in them. Risk analysis techniques characterize undesired outcomes and the uncertainties that surround them and estimate their probabilities of occurrence. Various techniques based in economics quantify outcomes in terms of costs and benefits, compare outcomes occurring at different times in the future, and aggregate the outcomes facing individuals into measures of net societal outcome. Spatial analysis and mapping techniques combined with ground-based or remote observation represent environmental change and its effects.

Such analytical tools help address several of the challenges of environmental decision making, but not all. In particular, they often fail to meet the challenges of value conflict and uncertainty. Value choices are often hidden in the simplifying assumptions of analytic techniques, and the assumed values may not be universally shared. Overreliance on analytic tools without adequate consideration of their limiting assumptions can sometimes heighten mistrust of governments and their experts and make it difficult to get public acceptance of public policy decisions.

Another problem that often arises with environmental analysis is a failure to address key decision-relevant questions. For example, a billion-dollar research program to assess the cancer risks of dioxin not only failed to resolve the scientific issues but also may not have been asking the right question, which, for many affected people, concerned the overall health risks to groups exposed to multiple hazardous chemicals, not just the cancer risks of dioxin exposure. In all the likely decision contexts in which these risks matter, dioxin is only one among many hazardous chemicals involved and cancer is only part of the problem (National Research Council, 1996). Several billions were spent to characterize the risks of leakage of radioactive materials from proposed repository sites for high-level radioactive waste, but no comprehensive appraisal was done to compare these risks with the risks of continuing current practices of temporary waste disposal (National Research Council, 1995; 2001a). Assessments of acid precipitation, including the U.S. National Acid Precipitation Action Program and similar assessments in several other countries, have been criticized for overemphasizing the collection of new data and not doing enough to interpret existing data to understand their implications for societal decisions (U.S. Office of Science and Technology Policy, 1983; U.S. Office of Technology Assessment, 1984; Oversight Review Board, 1991; Cowling, 1992;

Cowling and Nilsson, 1995). In short, when science is gathered to inform environmental decisions, it is often not the right science. Among the consequences are heightened social conflict, delayed decisions, and mistrust.

Because of these pervasive problems at the junction of environmental analysis and decision making, several authoritative studies have recommended processes that integrate analysis with broadly based deliberative processes involving the range of parties interested in or affected by the decisions (e.g., National Research Council, 1996; Presidential-Congressional Commission on Risk, 1997; Canadian Standards Association, 1997; Royal [UK] Commission on Environmental Protection, 1998). The goal is to put analysis more directly into the service of those who may be affected by decisions. In these analytic-deliberative processes, participants with diverse perspectives and values contribute to decision making in many ways, including defining the environmental decisions that require analysis, framing the scientific analyses needed to gain insight into the decisions, and interpreting the results to illuminate the decisions at hand (National Research Council, 1996).

Public agencies in the United States and elsewhere are increasingly committed to an analytic-deliberative approach to environmental understanding. For example, EPA has made extensive efforts to implement and improve "science-based environmental stakeholder processes" in support of its decisions (U.S. Environmental Protection Agency, 2001). The U.S. Climate Change Science Program, possibly the largest single environmental science program in the federal government and one involving 13 federal agencies, adopted "decision support" as one of four core approaches to meeting the program's goals in its current strategic plan (U.S. Climate Change Science Program, 2003). Among the implications of this emphasis are "early and continuing involvement of stakeholders . . . in defining key science and observation questions" and "transparent public review of analysis questions, methods, and draft results" (pp. 111-112).

Good environmental decision making requires both improved understanding of human-environment interactions and improved understanding and management of decision-making processes. The research we recommend will complement past efforts in the former area with expanded effort in the latter, using a decision science approach. Its central purpose is to identify and continually improve techniques for guiding and organizing practical environmental decision processes so that they achieve more of the ideal qualities of good decisions.

WHAT IS A DECISION SCIENCE APPROACH?

A decision science approach analyzes decisions and the processes for making them (e.g., Raiffa, 1968; von Winterfeldt and Edwards, 1986;

Morgan and Henrion, 1990; Keeney and Raiffa, 1993; Kleindorfer, Kunreuther, and Shoemaker, 1993; Clemen, 1996). It considers the objectives of decision making and ways to evaluate decisions and decision processes against those objectives. It identifies the questions and kinds of information needed for good decision processes, and it develops knowledge about which decision processes are likely to produce desired outcomes.

Basic research in decision sciences has focused primarily on unitary decision makers. Normative decision theory concerns hypothetical rational decision makers; behavioral decision theory concerns actual individuals (see the more detailed discussion in Appendix B). A smaller body of research has addressed the additional complications that arise when decisions are made in groups or when they must take into account the different and often conflicting perspectives and values of the people a decision will affect. There is relevant work in the social psychology of small-group decisions (Levine and Moreland, 1998) and in studies of organizational decisions, mainly in business (March, 1997). There are also numerous case studies of actual environmental decision making, but few of these have systematically applied conceptual frameworks (e.g., Renn, Webler, and Wiedemann, 1995; Beierle and Cayford, 2002).

This priority emphasizes research on what has been called prescriptive decision making (Bell, Raiffa, and Tversky, 1988), which we define as a science-based practice concerned with helping people to make good decisions. As already noted, our focus on the prescriptive presumes the continuation of basic research efforts on decision making that provide an essential foundation for prescriptive research. A decision science approach to practical decision making begins by identifying the elements of a responsible and competent decision-making process. For example, an ideal decision process has been defined on normative grounds as one that includes the elements listed in Box 2-1. These elements include some that are strongly dependent on participants' values (V) and some for which information from scientific and technical analysis (T) can provide participants with essential insight regardless of their values. Actual decisions vary widely in terms of how closely they approach these ideals.

Many analysts argue that environmental decision processes should be iterative to accommodate changing human desires and the changing state of knowledge about the effects of environmental choices. This is the notion of learning over time embodied in concepts of adaptive management and governance (Holling, 1978; Gunderson, Holling, and Light, 1995; Lee, 1993, Dietz et al., 2003; National Research Council, 1999a, 2004b).

From a decision science standpoint, good environmental decisions consider both physical and social phenomena—environmental processes, the available options, the effects of different options on environmental and social conditions, and so forth—and human values. Information about

BOX 2-1
Characteristics of an Ideal Decision Process According to Normative Decision Theory

- It clearly defines the decision to be made (V,T).
- It considers all the objectives that matter to the decision maker in making the decision (V).
- It identifies or creates a set of attractive alternatives for the decision (T,V).
- It considers the consequences of the alternatives in light of the available evidence (T).
- It considers uncertainties and unknowns regarding the consequences (T).
- It identifies and considers preferences regarding the trade-offs among the consequences of the alternatives (V).
- It selects alternative(s) on the basis of information about their consequences and the decision maker's preferences (V).
- It considers implications for linked and future decisions (T,V).

V = Achieving the ideal is strongly dependent on incorporating participants' values.
T = Scientific and technical analysis provide essential insight for achieving the ideal.
SOURCE: Adapted from Hammond, Keeney, and Raiffa (1999).

phenomena is often obtained from environmental scientists, health specialists, engineers, economists, and other experts on those phenomena. Information on values can legitimately come from a wide of array of interested parties. The need for both kinds of information is worth underlining because judgments about phenomena and about values are often intertwined, not least in the minds of analysts.

Good decisions require competent and socially acceptable ways to integrate information about phenomena with information about values (Hammond, Keeney, and Raiffa, 1999; Payne, Bettman, and Johnson, 1993; Kleindorfer et al., 1993). Decision science has developed some systematic techniques for doing this integration in ways that can be applied to environmental decision processes (Slovic and Gregory, 1999).

As already noted, environmental policy additionally involves diverse, conflicting, and changing values; substantial scientific uncertainty and ignorance; and often mistrust among participants. Consequently, judgments about information can be hotly contested. There may be disagreement about which technical information is needed and what its practical significance is, how to interpret uncertain or disputed information, how to make tradeoffs between desired outcomes, whether seeking more information will be worth the cost, and even about the nature of the decision to be made (National Research Council, 1994a, 1996:Chapter 2).

Decision science can help improve decision processes by making these judgments more explicit and structuring the ways participants in decision processes make and consider their judgments. It can also help parse disagreements so that decision participants can distinguish those that might be resolved by more information and clear thinking from those that may also require bargaining, compromise, or other resolution strategies. Decision science approaches have long been applied to a variety of environmental decisions (e.g., U.S. Nuclear Regulatory Commission, 1975; Lewis et al., 1975; Crouch and Wilson, 1982; Travis, 1988; Cohrsen and Covello, 1989; Rodricks, 1992; Suter, 1993) and used in training environmental policy analysts (e.g., Morgan and Henrion, 1990; Clemen, 1996). For further discussion of its use in environmental policy, see National Research Council (1996, 2002d) and U.S. Office of Management and Budget (2003).

Box 2-2 lists some characteristics of good public-sector environmental decision processes that have been proposed on the basis of previous research. These aspects of decision process are considered important for several reasons: normative (they allow affected parties to exercise democratic rights), substantive (they generate better alternatives and choices), and instrumental (they increase the likelihood of timely implementation) (Fiorino, 1990).

A major challenge in applying a decision science approach to environmental decisions is the linked nature of these decisions. Prescriptive decision theory will need to be expanded from its past emphasis on one-time

BOX 2-2
Some Proposed Characteristics of a Good Public Decision Process

- It appropriately represents the knowledge and perspectives of the spectrum of interested and affected parties to the decision.
- It ensures that each of these parties has sufficient access to expertise to allow meaningful participation.
- It uses a broadly based deliberative process to guide analysis so that technical information addresses the questions of greatest importance to the parties.
- It relies on information and analysis that meet high technical standards.
- It explicitly addresses scientific disagreements and scientific ignorance.
- It allows for reconsideration of choices in response to new information or changing values.

SOURCES: Webler (1995); National Research Council (1996).

decisions with defined boundaries to decisions that require linkages among administrative or institutional levels that have implications at many physical scales, that have long time horizons, and that involve iteration. The application of decision science to problems of long-term adaptive management or governance is an important area for contributions. At the other extreme, applying decision science to time-pressured decisions, as during crisis, also presents an important challenge and opportunity.

AREAS OF RESEARCH

Developing Criteria of Decision Quality

The quality of an environmentally significant decision cannot appropriately be defined by its outcomes because those outcomes may be highly dependent on factors that are unknown or uncontrollable when the decision is made. A decision may be well informed and well considered given what is known but may lead to unfortunate results because the system was not fully understood or because of the outcomes of key uncertainties. Thus, it is necessary to have internal criteria for judging the quality of decisions—criteria that are not dependent on ultimate outcomes. It is reasonable to expect that, on average, higher quality decisions (those based on the best available information, careful evaluation, adequate consideration of uncertainties and plausible worst cases, and so forth) are more likely than lower quality decisions to lead to desired outcomes in an uncertain world. It also seems reasonable to expect that higher quality decisions will on average be more widely accepted. There is evidence, however, from energy decision making of a perverse and inverse relationship between the quality of decisions and their acceptance (Craig, Gadgil, and Coomey, 2002). The empirical relationships among different indicators of decision quality is a worthwhile research question.

Researchers have proposed numerous internal criteria for decision quality. Some of these are presented in Boxes 2-1 and 2-2. The problem of defining decision quality for practical environmental decisions, however, has not received the level of research attention it deserves. Both the normative and the behavioral traditions in decision science have difficulty with this problem. The main difficulty in applying normative decision theory is that, in realistic situations, one cannot assess every consequence of every possible alternative and evaluate them all against the values of each decision participant. Decision makers seek the best practice not in the abstract, but under constraints of real people's cognitive capabilities, legislative mandates, limited time and resources, social conflict, and so forth: choices must be made and defended regarding what to include and what to exclude (for further discussion, see Clemen, 1996; National Research Council, 1996;

U.S. Office of Management and Budget, 2003). Researchers in the behavioral tradition typically do not address issues of decision quality, in part because their approaches emphasize characterization rather than evaluation and improvement.

The recommended research would build on recent efforts to develop empirically supported knowledge about ways to organize high-quality decision-making processes under realistic constraints (Renn et al., 1995; National Research Council, 1996; Beierle and Cayford, 2002; Webler, Tuler, and Krueger, 2002; Renn, 2003). It would inform the design of environmental decision processes by addressing questions such as these:

- What criteria do people use to evaluate decision quality? To what extent do these criteria differ for people from different cultural, socioeconomic, or educational backgrounds or for people representing different positions in environmental controversies? To what extent do they depend on past experience with related decisions?
- Which characteristics of decision processes are associated with judgments of decision quality by the participants or outside observers? Which characteristics are associated with confidence in decisions? With acceptance of decisions?
- How do different ways of organizing decision processes fare in terms of the attention given to the elements of normatively good decision process (for example, do processes that involve more different stakeholders do a better job of identifying all relevant objectives, as is often claimed)?
- How do different levels of attention to particular elements of good decision process affect assessments of overall decision quality?
- How do different resolutions of trade-offs in the decision process (e.g., between getting more information and deciding quickly, or between broader representation and efficiency of decision) affect various indicators of decision quality?
- To what extent do interventions designed to ensure that decisions address certain elements of good decision processes result in more positive assessments of decision quality? Which elements are most important to those judgments under which conditions?
- When decisions are highly constrained in terms of time, attention, or legal requirements, which elements of good public decision processes are most critical to the quality of the decisions?
- Are decisions of higher normative quality associated with preferred social and environmental outcomes?
- How can research results concerning good decision processes and ways to promote them best be disseminated to the users of these results (e.g., government agencies, stakeholder groups, corporations, partnership groups)?

This research might be conducted by various methods, including structured comparisons of naturally occurring cases, simulation, modeling, and quasi-experimental field research. Because environmental decisions present the full range of difficulties in decision making outlined above, they provide a good test bed for research on decision quality more generally.

Developing Formal Tools for Structuring Decision Processes

Behavioral decision research shows that individual decision makers typically omit key elements of good decision processes and that their decisions suffer as a result (Slovic, Fischhoff, and Lichtenstein, 1977). People respond to complex tasks by using their judgmental instincts to simplify them in ways that seem adequate to the problem at hand. They respond to probabilistic information or questions involving uncertainty with predictable biases that often ignore or distort important information (Kahneman, Slovic, and Tversky, 1982). They have difficulty clarifying objectives (March, 1978), identifying all viable alternatives (Keeney, 1992), and structuring decision tasks (Simon, 1990). When asked to consider value tradeoffs or select among alternatives, they employ heuristic reasoning processes that are susceptible to a variety of contextual or task-related influences (Payne et al., 1993). Hence, there are many reasons to expect that, on their own, individuals (including experts) will often fall short of the normative ideal in making choices about complex issues involving uncertainties and value trade-offs.

Decision-making groups can, in principle, identify more elements of any decision than individuals and correct for individual members' errors, thus producing better decisions than individuals. Behavioral research on group decision processes indicates, however, that this potential is not necessarily realized in practice. Individuals who have relevant information that is not widely shared in the group must get others to take this information seriously. Whether this happens is highly contingent, depending for instance on the individual's social status (Hastie et al., 1983; David and Turner, 1996), the group's norms (e.g., a value on originality opens the group to individuals' information, a value on agreement closes it; e.g., Moscovici, 1985), and the tactics the individual uses to propose the ideas (Turner, 1991). Moreover, social processes in groups sometimes lead to premature closure on a common perspective that ignores contrary information, resulting in tendencies toward "group think" (Janis and Mann, 1977) or group polarization (Kaplan and Miller, 1987).

The decision sciences have developed a variety of tools to structure decisions and help decision makers and decision-making groups better approximate ideals of good decision processes. The recommended research will refine these tools and apply them more widely as a basis for improving environmental decisions. To illustrate the possibilities, we briefly describe progress on developing tools for three purposes: clarifying decision participants' values and preferences concerning alternatives, understanding and thinking through uncertainties and disagreements about the implications of choice options, and assisting in making collective choices when different individuals have conflicting understandings and competing preferences.

Clarifying Values and Preferences

The values and preferences people express regarding complex and unfamiliar environmental goods vary considerably according to how they are elicited (Payne et al., 1993; Slovic, 1995). Hence, the task of clarifying preferences is one of helping people construct their preferences rather than simply revealing them. Tools concerned with preference construction and elicitation include formal methods based on precepts of measurement theory and decision theory and wisdom gleaned from applied experience. Perhaps the most well-known of these formal approaches is termed multiattribute trade-off analysis, which involves an interview between an analyst and a decision participant (Keeney, 1992). The result is a mathematical statement comprising a utility or value function that could be used to evaluate every possible alternative within the range of consequences used in the interview process.

The advantages of formal techniques of preference construction are that the judgments involved are made explicit, that the value information can be used in many ways to help clarify the decision process, and that decision makers in collective choice situations can learn a great deal through joint efforts to clarify preferences. The disadvantages are also substantial: the questions involved are difficult to answer and require decision makers to make their inchoate feelings explicit, the questioning process may be confusing, the process can be cognitively and analytically demanding, and it may not be clear how the results will be used. This approach has other drawbacks, including the lack of trained people to implement such preference elicitation approaches, and the lack of a rigorous way to combine individuals' utility functions into guidance for collective decisions, such as the kind of social welfare function provided by welfare economics.

In part as a response to these drawbacks, several other approaches to preference elicitation have been developed and tested by researchers in order to make the task cognitively simpler, more transparent, or more closely matched to particular decisions. These include the analytic hierarchy process developed by Saaty (1980, 1991); strategies that focus on a choice among a set of possible policy alternatives to address a given environmental question (McDaniels and Thomas, 1999); methods of clarifying preferences based on judgments of what would constitute "even swaps"

(Hammond, Keeney, and Raiffa, 1999); and techniques that emphasize key aspects of the decision problem, such as values ("value-focused thinking," Keeney, 1992), particular objectives, or finding an alternative that provides acceptable performance across all the objectives (satisficing) (Payne et al., 1993). Such approaches are discussed in more detail in Appendix B.

Understanding Uncertainties and Disagreements

Decision scientists have been developing analytical tools and approaches for characterizing uncertainties. These include methods of eliciting and making use of probabilistic judgments or other sources of information about uncertainty, methods of combining probabilistic estimates through simulation, and methods of characterizing several different sources of uncertainty at once, all as a basis for estimating the effects of decision options (Morgan and Henrion, 1990; Cullen and Frey, 1999). Decision researchers have also experimented with methods of conveying information from these methods to decision participants as a basis for better understanding, deliberation, and decision making. Morgan and Henrion (1990) provide a comprehensive review of such methods and how they are applied for complex problems.

Although probability estimation remains the major approach to characterizing uncertainties, other methods are being developed that involve less demanding judgments. Some are based on fuzzy set theory (Zadeh, 1965), on the presumption that highly precise probabilistic judgments are often unnecessary. Scenarios offer another widely applied approach to characterizing uncertainties for environmental decisions (Waack, 1985a, 1985b), although relatively little research has been conducted on their efficacy as a means to generate an appropriate comprehension of uncertainty (see Chapter 6; Moss and Schneider, 2000).

Recent research using influence diagrams (a generic tool with wide application in model building, problem structuring, probability elicitation, knowledge mapping, and many other contexts) helps reveal the mental models of decision participants and the sources of some of their disagreements (Howard, 1989; Clemen, 1996). Influence diagrams can reveal differences in the understandings of lay and expert participants or between participants with different stakes in the decision (Morgan, Fischhoff, Bostrom, and Atman, 2002). They can help participants understand the bases of disagreements and perhaps see ways to resolve them.

Assisting in Collective Choice

One of the most difficult challenges in environmental decision making is how to arrive at a societal preference in a collective decision context.

Since the writing of Arrow (1963), decision researchers have recognized that there is no unique rule for aggregating the ordinal preferences of individuals with different values across a range of alternatives. Research is warranted on a variety of techniques that may be useful for informing judgments about societal preference. Voting approaches could possibly provide a means of directly eliciting preferences and under specific rules provide affording a basis for aggregation, although such approaches are highly affected by how questions are framed, the set of alternatives, and the choice of aggregation rules (Brams and Fishburn, 2002).

New information technologies may also provide useful tools for expression of individual preference, even if a generalized rule for social choice on the basis of expressed ordinal preference may be impossible. Computer-based tools for knowledge and value elicitation may provide widely applicable approaches to obtaining high-quality judgments about subjective probabilities of consequences and the values people associate with different consequences. Problem-structuring tools such as influence diagrams may have enormous potential in conjunction with advanced information technologies.

We are not recommending new research related to benefit-cost analysis, even though this approach is widely used to address the key issue of arriving at a social choice. We have two reasons for not doing so. One is that this line of research has its own momentum and seems less in need of increased research support than other, less developed areas. The current state of concepts and practice for benefit-cost analysis are discussed in several sources (e.g., Cropper and Oates, 1992; Zerbe and Dively, 1994; Morgenstern, 1997; Boardman, Greenberg, Vining, and Weimer, 2000; U.S. Office of Management and Budget, 1996; and Freeman, 1993).

The other is that benefit-cost analysis is in some respects antithetical to the research program recommended here because it decides by assumption how to address at least two important issues that we think need to be decided empirically. It assumes that social value is nothing more or less than the sum of the values individuals express in markets or market-like contexts, and it assumes that the values of different kinds of consequences (for employment, endangered species, sacred spaces, etc.) can be compared by reducing them to a single monetary metric.

There is growing literature documenting difficulties with these assumptions (Kelman, 1981; Morgan, Kandlikar, Risbey, and Dowlatabadi, 1999; Lave, 1996), particularly for large-scale problems involving long time horizons, nonmarginal changes, deeply held values, and equity issues. Moreover, because these assumptions are sometimes not shared by people affected by environmental decisions, attempts to employ them on actual environmental policy decisions have proved controversial and divisive (National Research Council, 1989, 1996). A large literature on perceptions of justice and injustice, although not directly addressed to environmental issues, makes it clear that for people in the United States and several other countries, concepts of just decisions do not reduce to the choice that is best for the individual making the judgment and that individuals' normative judgments about whether decisions are just can engender predictable emotional reactions (e.g., anger, resentment) that it may be risky for collective decision processes to ignore (see, e.g., Tyler and Smith, 1998; Miller, 2001; Mikula, 2003; Skitka and Crosby, 2003).

The research recommended here would investigate ways to structure decision processes, develop empirical understanding of the effects of various decision rules and analytical assumptions, and identify ways to structure decisions that help actual decision processes more closely approach normative ideas of good decision making. Research on formal tools for structuring decision processes might address such questions as these:

- How can formal methods for improving decisions be made understandable and cognitively tractable for participants in complex environmental decisions? How can such methods be applied in real-world decision settings? How are the decisions affected?
- To what extent and under what conditions do the benefits of formal approaches to decision making outweigh their costs in time, money, and effort?
- How can judgments about the nature and likelihood of a range of outcomes be made more routine and workable through the use of information technologies? Do approaches such as influence diagrams and elicitation of subjective probability lead to clearer and more accurate understanding of uncertainty?
- How can learning be built into these formal tools through the potential for updating over time?
- How can methods for structuring decisions be applied effectively when decision processes overlap and involve multiple agencies, levels of organization, and sectors that jointly affect environmental outcomes?
- What systematic methods for aggregation of preferences can be developed and implemented in realistic environmental decision settings that do not require the strict assumptions of social benefit-cost analysis?
- How can risk communication methods be used to make the results of efforts to clarify preferences and uncertainty intelligible and useful to key decision makers and affected parties?
- Which values matter to individuals in important generic decision situations (e.g., purchase of energy services, housing, transportation, and consumer durables)? Can decision-aiding approaches help consumers by structuring the values and uncertainties in these choices as well as their links to other broader level decisions?

Creating Effective Analytic-Deliberative Processes

As already noted, several authoritative studies recommend that public policy decisions affecting environmental and associated public health risks be organized in ways that integrate analysis with broadly based deliberative processes involving the range of parties interested in or affected by the decisions. These studies conclude that better decisions can result when analysis is organized for decision relevance by giving decision participants a guiding role: "deliberation frames analysis, analysis informs deliberation, and the process benefits from feedback between the two" (National Research Council, 1996:6).

Many government agencies in the United States and elsewhere have made commitments to using broadly participatory processes involving analysis and deliberation to make or support environmental policy decisions and many have tried to implement those commitments (see, e.g., Beierle and Cayford, 2002; Leach, Pelkey, and Sabatier, 2002; Bradbury et al., 2003; Kasemir et al., 2003). Nevertheless, the quality of these decisions is only beginning to be evaluated and the knowledge base for selecting the best process for a specific decision remains weak. By the late 1990s it was possible to demonstrate the potential of analytic deliberation, to identify some of the factors likely to affect its success, and to show that the best process depends on the situation. But because systematic analyses based on data from multiple cases are only beginning to appear (Jones and Klein, 1999; Beierle and Cayford, 2002; Leach et al., 2002; Bradbury, Branch, and Malone, 2003) and because most of these studies are restricted to specific decision contexts, understanding has not progressed to the point at which science-based input can be given to the design of processes affecting types of decision that have not yet been studied. As a result, organizations that convene such processes have been limited to improvising on the basis of the judgments of experienced practitioners and extrapolation from available case studies.

This situation is ripe for change. In recent years, researchers have begun to apply consistent methods to the study of multiple analytic-deliberative processes (e.g., Ashford and Rest, 1999; Beierle and Cayford, 2002; Leach et al., 2002; Bradbury et al., 2003). Such studies have the potential to demonstrate generalities that apply across contexts and to specify ways in which outcomes are context-dependent. These studies, together with advances in theory and conceptualization (e.g., Renn, Webler, and Wiedemann, 1995; National Research Council, 1996; Beierle and Cayford, 2002; Renn, 2003), are making it possible to build much more nuanced understanding of desired outcomes, such as how decision quality and legitimacy are affected by the ways collective environmental decision-making processes are organized (e.g., whether and how the parties are represented, what resources they have available, how their input is structured, how decision makers are constrained in using external input).

With this base of concepts and empirical knowledge, researchers are now poised to draw on preexisting bodies of basic social and behavioral science research that are clearly relevant to the design of environmental decision processes. These include not only decision research, as already noted, but research on small-group processes (Moscovici, 1985; Levine and Moreland, 1998; Mendelberg, 2002), perceptions of justice and fairness (e.g., Tyler and Smith, 1998; Mikula, 2003), democratic deliberation and civic participation (e.g., Fishkin, 1991; Elster, 1998; Dryzek, 2000; Ostrom, 1990); organizational change (e.g., Scott, 1992; Chess, 1999), communications research (National Research Council, 1989; McComas, 2001), and conflict resolution (e.g., Druckman, Broome, and Korper, 1988; Rubin, Pruitt, and Kim, 1994; Fisher, 1997; Bingham and Langstaff, 2003). With clearer conceptual frameworks for examining environmental decisions, findings from these separate, older lines of research can be linked to the study of environmental decisions and can generate new and fruitful hypotheses to explore.

An ongoing study on public participation in environmental assessment and decision making at the National Research Council is synthesizing knowledge in this rapidly moving field and the associated fields of basic social and behavioral science and developing recommendations for research and practice (see http://www7.nationalacademies.org/hdgc/Public_Participation.html). An organized research community is beginning to emerge that can generate the knowledge needed for a science-based practice of process design for environmental policy decisions. This research can improve the ability of decision-making organizations to deal in a competent and credible way with environmental complexity, incomplete and uncertain knowledge, diversity of human values and interests, and the other realities that make this field of decision making so difficult.

Research to build effective analytic-deliberative processes could address such questions as:

- What are good indicators for key attributes of success for analytic-deliberative processes, such as decision quality, legitimacy, and improved decision capacity?
- How are these outcomes affected by the ways in which the processes are organized, the range and diversity of people involved, the rules used for deliberating and reaching conclusions, the ways technical information is organized and made available, and the environmental, social, organizational, and legal contexts of the decision at hand?
- What are effective ways to make technical analyses transparent

to a wide range of decision participants, some of whom lack technical training?

- How can decision-analytic techniques for preference elicitation, characterizing uncertainty, and aggregating preferences be used to best advantage in broadly based analytic-deliberative processes?
- How can decision processes be organized to ensure that all sources of relevant information, including the local knowledge of nonscientists, are gathered and appropriately considered?
- How can analytic-deliberative decision processes be organized to reach closure effectively and with broad acceptance, especially when the processes involve a diversity of perspectives and interests? What tests could be applied to decisions and decision processes to support claims that they are ready for closure?

RATIONALE FOR THE SCIENCE PRIORITY

Likelihood of scientific advances. The recommended research can yield significant scientific advances by building on several recent developments in understanding. Recent efforts to identify and assess several elements of decision quality (e.g., Webler et al., 2002; Beierle and Cayford, 2002) have established the groundwork for much improved understanding and measurement of this concept. Substantial recent work on decision-analytic tools for structuring decision processes, conducted mainly in laboratory and simulation settings, provides a basis for developing these tools further and testing and comparing their usefulness in realistic settings involving multiple and diverse participants. Recent theoretical, conceptual, and empirical work on analytic-deliberative processes and the increasing development of a self-identifying community of researchers and practitioners has set the stage for rapid progress through conceptually coherent empirical research on the design and study of processes for informing environmental decisions through analytic deliberation.

Continuing interest at the National Science Foundation in research on environment and decision making bodes well for scientific advances. Through its programs on decision, risk, and management science and human dimensions of global change and its initiatives on coupled human and natural systems and environmental research and education, the foundation is bringing together researchers from multiple disciplines with shared interests in environmental decision processes. These venues for communication are likely to provide good test beds for new research ideas.

Potential value. The long history of inadequately informed and incompletely deliberated environmental decisions, as well as the cost in delay, decisional gridlock, social conflict, and mistrust of government, make clear the importance and value of finding more competent and legitimate ways to organize the processes that lead to public policy decisions affecting environmental quality. Moreover, the increasingly widespread practice among federal agencies and other governmental and nongovernmental entities of opening environmental decision-making processes to a range of stakeholders and potentially affected parties has raised the stakes for managing decision-making processes well. A decision science approach can increase the likelihood of success with such processes.

Likelihood of use. The increasingly widespread use in government of participatory processes requiring both analysis and broadly based deliberation indicates the potential demand for scientifically informed guidance on how to make decision processes work better. Despite the public commitments of various government agencies to openness, however, significant barriers remain to the use of results from the recommended scientific research on decision-making processes. These include commitment to standard procedures or past practices, perceptions of statutory constraints, and a shortage of organizational capability to conform to principles of sound process design (National Research Council, 1996). Key decision makers may not recognize that it is possible to put the design of decision processes on a scientific footing. Perhaps the most serious barrier to use of the results of the recommended research lies in the potential unwillingness of some decision makers to delegate responsibility for the design of decision processes or to involve a full range of affected parties in decision making in a serious way for fear that the ultimate decision might not fit their preconceived ideas or serve interests they wish to promote.

Despite such potential barriers, many environmental agencies clearly have backed their stated commitments to better and more open decision processes with significant investments of time, money, and institutional reputation, for example, in seeking out and responding to the input of a variety of stakeholders in these processes. Some have also shown serious interest in designing these processes with the help of sound knowledge. Such agencies are likely to take research results seriously if researchers are given incentives to disseminate their findings and if good lines of communication are established between researchers and practitioners. To make best use of research results, decision-making organizations should create internal incentives and assign responsibility within the organization for incorporating the best science into the deliberative part of decision making, and not only the analytical part. To the extent that these efforts are successful in some public-sector organizations, they are likely to diffuse to others over time.

3

Institutions for Environmental Governance

Federal scientific and environmental agencies should support a concerted effort to build scientific understanding needed for designing and evaluating institutions for governing human activities that affect environmental resources. Environmental governance refers to any institutional arrangement that attempts to control individual or organizational use of natural resources, ecological systems, and sinks for wastes in order to meet objectives such as sustainable use, protection of public health, and protection of valued species or places. Societies have developed many institutional structures for environmental governance, all of which are effective in some circumstances, but none of which is universally successful. This priority is to build more systematically the knowledge needed to design institutional forms, that is, sets of rules and associated cultural and organizational systems, that can effectively address specific environmental governance problems. In identifying this priority for research, we concur with previous reviews that have identified the same area as a "research imperative" (National Research Council, 1999b), a "grand challenge" in environmental sciences (National Research Council, 2001b), and a major research challenge in environmental research (Pfirman and the NSF Advisory Committee for Environmental Research and Education, 2003). The area presents challenges, but it is also ripe for progress.

THE RESEARCH NEED

Environmental resources present the governance problems typical of common-pool resources, that is, resources that can be used simultaneously by more than one user, and anyone's use potentially degrades the resource for all. When ownership and the assignment of rights and responsibilities are unclear and access is unrestricted, these resources often generate so-called social dilemmas or social traps in which the outcome of decision making is less than optimal, if not wasteful or destructive. Part of the difficulty is a fundamental issue in social interaction, the free-rider problem, which arises when the results of coordinated social action are public goods, available to everyone, so that the incentive is reduced for any user to contribute to their management (Olson, 1965). Without effective rules restricting access, even evidence of resource decline may fail to induce restraint, resulting in a "tragedy of the commons" (Hardin, 1968). Tragic outcomes are especially difficult to avoid when resources are highly unpredictable and poorly understood (Wilson, 2002).

A major finding of recent decades of research is that such results can be avoided through institutions for governing the commons that meet basic requirements of environmental governance, such as providing needed information and infrastructure, resolving conflict, inducing compliance with rules, and adapting to change (e.g., Ostrom, 1990; National Research Council, 2002a; Dietz, Ostrom, and Stern, 2003). By institutions we refer to rules and the social and cultural systems that maintain them. Common institutional forms include direct control by centralized government agencies; indirect control through quasi-privatized and tradable allowances or quotas; nongovernmental control through market mechanisms; nongovernmental control by associations of businesses, communities, and resource users or by representation of diverse interests on decision-making bodies (e.g., of environmental interests on corporate boards); partnerships and collaborations that cross jurisdictional or sectoral lines; and participatory forms of governance that combine expert and lay knowledge and authority. Although each of these institutional forms can meet governance requirements under the right conditions, none is uniformly successful. The research need is to develop sufficient knowledge to enable improved choices of institutional forms that are well suited to meeting environmental and other objectives in particular situations and at particular spatial and temporal scales, as well as being capable of adapting to the dynamics of complex socio-ecological systems.

Analyzing environmental governance as a problem of institutional design is useful because it reframes the central governance question from one of selecting a single best governance strategy (e.g., choosing between top-down regulation and market-oriented policies) to one that considers a full range of governance options and seeks to match institutional forms to specific governance needs. It expands discussion from a debate over which actors are best able to govern resource use (e.g., national governments versus local governments, governments versus businesses) to a discussion of the most appropriate roles in governance systems for all types of actors, including governments, businesses, formal and informal "civil society" organizations, scientific groups, and individuals.

Environmental policy in the United States has been moving slowly toward recognition that a broad array of governance options is available. At first, policies were commonly built on assumptions about the authority and legitimacy of centralized government (the state), the centrality of science, and the possibility of full understanding and control of natural and social systems. From the turn of the twentieth century, a dominant and continuing model for environmental policy was the multipurpose management of public lands, forests, surface waters, wildlife, and minerals underlying public lands and waters by federal agencies, based on scientific and technical expertise and guided by a doctrine of public trust to serve the overall public interest. Beginning in the New Deal era, this model was augmented by the large-scale use of government subsidies, especially for multipurpose water management projects, and by partnership arrangements with favored resource user interests, such as farmers and ranchers. In the late 20th century, it was supplemented by an unprecedented new suite of federal regulatory statutes mandating control of pollution and of toxic contamination, as well as by vastly increased federal subsidies for wastewater treatment facilities and for cleanup of sites contaminated by toxic chemicals (Andrews, 1999). Difficulties with these forms of governance have contributed to challenges to them and to their underlying assumptions, leading in some cases to the introduction of more market-oriented mechanisms for compliance, such as tradable property rights, information disclosure requirements, and strict liability principles, as well as statutorily automated and nondiscretionary penalties for noncompliance.

The question of environmental governance is typically posed as a choice among a few basic policy strategies. The dominant strategy—direct management or regulation by centralized government agencies, with top-down creation and implementation of rules imposed on businesses and utilities—has generated considerable dissatisfaction in the United States in recent years. Rules derived from national legislation such as the Clean Air Act, the Clean Water Act, and the Comprehensive Environmental Response, Compensation, and Liability Act (Superfund) have contributed to major reductions in pollution discharges and environmental contamination hazards. They also have been criticized for varying combinations of administrative burdens, technological rigidity, imperfect compliance, imposition of uniform national approaches on diverse environmental circumstances, political influence on the regulatory process by regulated interests, and regulation of some source categories but not others; and, as a result, for gaps between outcomes and legislative objectives (Vig and Kraft, 2003). Similar mixtures of success, criticism, and political conflict have been directed at top-down natural resource management programs (forests, grazing lands, fisheries, wildlife) in the United States (Knight and Bates, 1995; Weber, 2002) and elsewhere, particularly developing nations (Gibson, McKean, and Ostrom, 2000; Hulme and Murphree, 2000).

Criticism of centralized environmental management and regulation has drawn attention to an alternative institutional form—market-based governance—which has been advocated as more flexible and more economically efficient. This approach relies on creating incentives for individual and firm behavior by privatizing certain rights and allowing markets to emerge for them (Freeman, 2003). Examples include tradable environmental allowance schemes for regulating pollutants and individual fishery quota schemes for fisheries management. Government establishes a limit to resource use, allocates use rights, and allows those rights to be exchanged in markets in a "cap-and-trade" system, in which governments set limits on resource use or pollutant emissions, allocate initial rights to those resources or pollutants, and allow those rights to be traded in the market (Rose, 2002; Tietenberg, 2002; Young and McCay, 1995). Although this approach has performed well in some policy arenas (Tietenberg, 2002), it has been very controversial in general and in particular applications, and it has not always lived up to its advocates' expectations (see Marine Fish Conservation Network, 2004, and U.S. General Accounting Office, 2004, for the fishing case; Solomon and Lee, 2000, on sulfur emissions; Lee, 2004, on mercury emissions from coal-fired power plants; McCay and Brandt, 2001, on surf clam and ocean quahog quotas). Tradable allowances and related approaches are sometimes presented as an alternative to centralized government regulation and management, but in fact they combine centralized regulation with a market-based procedure for allocating a property right that has been created by central government.

Dissatisfaction with national-level control has also led to interest in approaches that decentralize or devolve elements of institutional authority and responsibility from the federal government to state or local governments, as for example, the National Environmental Performance Standards arrangements between the U.S. Environmental Protection Agency and many states created during the Clinton Administration (Rabe, 2003), or to firms or other private-sector organizations, as in so-called voluntary alternatives to regulation (National Research Council, 2002b; see also Chapter 4). The theme of devolution is also evident in natural resource policy (Lowry, 2003), particularly in the American West, where much of the land is under federal ownership and control (David, 1997; Steel, 1997).

Shifts from top-down, direct regulation by the federal government to other forms and levels of governance have been accompanied by interest in deliberative, discursive, and participatory approaches (Dryzek, 1990; Press, 1994; Renn, Webler, and Wiedemann, 1995; National Research Council,

1996) and improved processes of public participation (Kelleher and Reccia, 1998; Beierle and Cayford, 2002), especially when health risks are involved (Chess, Hance, and Gibson, 2000). These approaches are discussed in Chapter 2 in relation to the need for a science-based approach to developing participatory decision processes. They also play an important role in efforts to make environmental science increasingly decision relevant (see Chapter 6). The effectiveness of participatory approaches to environmental governance, particularly in natural resource management and other policy arenas that require a long series of decisions over time, depends also on the creation of organizations and rules that can maintain the quality of decision processes over time and induce compliance with decisions. In this context, an important institutional innovation is collaborative planning (Brick, Snow, and Van de Wetering, 2001; Porter and Salvesen, 1995) involving public-private partnerships and multistakeholder groups (e.g., Leach, Pelkey, and Sabatier, 2002). This approach faces challenges of implementation, particularly in developing institutional frameworks for improved communication and cooperation between technical and scientific experts and lay members of the public (Fischer, 2000; Irwin, 1995) and in creating community-based programs of environmental protection, restoration, and management that act consistently with the responsibilities of higher levels of government (Brick et al., 2001; U.S. Environmental Protection Agency, 1997, 2002a).

Attention to ideas of ecosystem management has added further complexities to questions about environmental governance, including how to deal with mismatches between the jurisdiction and scope of government and the spatial and temporal scales and dynamics of ecological and human-ecological systems (Lee, 1993) and how to address the stubborn persistence of certain long-lived institutional arrangements (Wilkinson, 1992). Concepts such as adaptive management, though appealing in principle, have proved difficult to implement, largely due to institutional problems (Gunderson, 1995; Walters, 1997).

Environmental governance institutions must increasingly deal with new challenges. For example, those affected by environmental decisions, even at the local scale, may have very heterogeneous backgrounds, needs, and interests to be represented and considered. Nongovernmental organizations (NGOs), international lending institutions, and private foundations sometimes strongly influence or overshadow government agencies (World Resources Institute, 2003; Ribot, 2002). For example, structural readjustment policies of the International Monetary Fund have led governments to cut back on their services while transnational and national NGOs were trying to help local communities exercise governance over local resources such as forests, waterholes, and fisheries. Privatization of certain common resources, such as drinking water, has sometimes been promoted by international agencies and by trade associations, while being contested by grassroots organizations and national and transnational NGOs.

The developments just discussed have led to increased attention to decentralization, pluralism, and innovation in crafting and adapting institutions for environmental and natural resource governance (Wilson, Nielsen, and Degnbol, 2003; National Research Council, 2002a; Schelhas, 2003; Haas, 2004a). They also raise questions about the appropriate roles of a variety of governmental, private-sector, and nongovernmental or civil society organizations in governance systems that involve all these types of participants. The critical research need is to develop the knowledge needed to inform decisions about how to choose effective institutional forms or develop new forms drawing on elements from existing approaches so as to create governance institutions that will work well for specific environmental governance problems.

AREAS OF RESEARCH

The recommended research would explore questions of environmental governance by linking traditional approaches to policy analysis and evaluation (discussed in more detail in Chapter 6 and Appendix D) with a research tradition that conceptualizes environmental governance more broadly in terms of institutional design. The institutional research tradition, which builds on theory in several social science fields, has broadened its scope over time to encompass a wide range of environmental and other resources. Research on institutions for managing common-pool resources (e.g., Ostrom, 1990; National Research Council, 2002a) has developed an intellectual framework that can help refine past debates about the relative merits of command-and-control, market-based, and voluntaristic policy strategies. Research can be focused on how particular institutional forms address the basic tasks of environmental governance in specific environmental and social contexts and to identify strategies, including combinations of institutional forms, that are likely to perform those tasks well in particular contexts. The research needs have been described in considerable detail elsewhere (National Research Council, 1999b, 2002a). Here we identify a few illustrative and promising areas of research.

Requirements of Governance

Recent reviews (National Research Council, 2002b; Dietz et al., 2003; Acheson, 2003) have identified key requirements for adaptive governance of complex systems of human-environment relationships. These requirements suggest questions for future research that can be pursued in studies of specific decisions at specific sites and in comparative research across settings aimed at building knowledge about ways to meet the governance requirements in specific settings and at particular scales. These questions include:

- What kinds of information are needed for effective governance and through which mechanisms or organizations can they most effectively be provided?
- What are the key conflict issues and the effective ways to manage them?
- What are the most promising strategies for inducing rule compliance?
- What kinds of physical and informational infrastructure are needed for governance, and how might they best be provided?
- What characteristics of governance institutions are most likely to enhance the capability to adapt effectively to change?
- In what ways are the answers to the above questions contingent on aspects of the environmental, social, political, and economic context?

Property Rules

Property rules are key to participation in governance, to conflict management and rule compliance, and to resource distribution. They assign rights to outputs of common resources and other matters, including decision-making rights and responsibilities. Research has helped refine understanding of property institutions from the simple distinction between public and private property to recognition of a broader variety of property regimes, including "no property" (open access) and common or communal property (Feeny, Berkes, McCay, and Acheson, 1990). It has recognized the complexity and plurality of property regimes (Geisler and Daneker, 2000) and their embeddedness in particular political, historical, and cultural systems (Hann, 1998; McCay, 2002; McCay and Acheson, 1987). These refinements have begun to be applied to the analysis of land use, environmental, and natural resource management questions in the United States (Cole, 2002; Geisler and Daneker, 2000). This research has included comparative institutional analyses that compare community-based management regimes with market-based property regimes, such as tradable environmental allowances (McCay, 2000; Rose, 2002). These analyses address a variety of issues, including environmental outcomes, economic efficiency, and social equity. They also consider the effects of ecological, technological, institutional, and cultural context and show why institutions that function consistently across settings in theory may in fact function quite differently in different settings. For example, tradable permit systems that look the same from a theoretical standpoint have performed much better in managing air pollution than in managing fisheries or fresh water supplies (Tietenberg, 2002). Future research on property institutions in their contexts can more fully illuminate choices of institutional forms to match their settings and identify opportunities for adaptation.

Legitimacy and Trust

Conflict management and rule compliance are affected by the heterogeneity of resource users, the kinds of communication and levels of trust among them, and the perceived legitimacy of governance institutions (Falk, Fehr, and Fischbacher, 2002; Kopelman, Weber, and Messick, 2002). Future research can usefully focus on how findings from experimental research on trust, reciprocity, and related aspects of decision processes relate to experiences in actual resource governance situations at different scales and on how various features of decision situations and institutions affect trust, communication, and legitimacy. Future research should also address how legitimacy and trust—and hence the effectiveness of commons institutions—are affected by the increased complexity and scale of environmental problems and interested publics.

Linkages Across Scales

The trend toward decentralization and devolution, increased interest in "co-management" institutions, and the need to govern transnational environmental resources all raise questions of how to integrate smaller-scale and place-based institutions with higher levels of governance (Hanna, Folke, and Maeler, 1996; McCay and Acheson, 1987; Ostrom, 1990; Wilson et al., 2003; World Resources Institute, 2003). Such cross-scale linkages are a critical focus for future research on institutional design. Vertical linkages between local-level institutions and subnational or national ones or between national and international ones, both governmental and nongovernmental, are often characterized by tensions and unintended consequences (Young, 2002). There are trade-offs between the potential benefits of higher level arrangements, such as efficiencies of scale, correspondence with large-scale ecological structures and functions, and avoidance of externalities problems, and the benefits of smaller scale institutions, such as more accurate monitoring of environmental variation and variability and the ability to use low-cost informal sanctions to induce compliance (Berkes, 2002; Wilson, 2002; Young, 2002). The increasing globalization and complexity of environmental problems and governance underline the importance of developing governance systems that cross scales, improve information flows, and allow for high levels of flexibility and adaptability (Cash and Moser,

2000; Haas, 2004a; Ostrom, 2001; Wilson, 2002). Developing and improving such systems is a very high priority for research.

Future research should address questions of scale linkage such as these:

- To what extent can lessons learned at one level of governance transfer to other levels?
- How can subnational institutions be effectively linked to management objectives established nationally or internationally (e.g., climate change)?
- How do the proliferation of horizontal linkages such as indigenous people's movements and the increased activity of nongovernmental organizations affect environmental governance?
- What are the comparative advantages in complex governance systems of the various governmental and nongovernmental actors, including NGOs, multinational corporations and business associations, scientific networks, and international institutions that have developed some autonomy?
- How can governance institutions be structured to provide for an effective division of labor among the above actors in meeting the requirements of environmental governance?

RATIONALE FOR THE SCIENCE PRIORITY

The rationale for devoting significant research efforts to understanding environmental governance in terms of institutional design has been laid out in several previous National Research Council reports, one of which placed the topic on a short list of grand challenges in environmental science (National Research Council, 1992, 1999b, 2001b) and in a recent priority-setting exercise at the National Science Foundation (Pfirman and the NSF Advisory Committee for Environmental Research and Education, 2003). Research on environmental governance continues to deserve high priority under the decision criteria imposed by this study.

Likelihood of Scientific Advances

Improvements over the past decade in the conceptual framework for understanding environmental governance and the development of growing data bases of comparable cases (see the common-pool resources database maintained at Indiana University, online at http://www.indiana.edu/~iascp/Iforms/searchcpr.html) have created a very favorable situation for scientific advance in the understanding of the functioning of systems for environmental governance. Conceptually guided case comparisons, experimental simulations, and modeling of governance systems together provide a very strong base for developing generalizations about the effects of particular institutional forms and for showing how and where these generalizations are context dependent. It is becoming possible to move beyond normative modes of analysis toward far more pragmatic approaches (see Haas, 2004b) that are capable of yielding helpful results.

Potential Value

The products of the recommended research can be of considerable value to a variety of environmental policy actors by providing them with a useful and flexible conceptual framework and a growing body of knowledge, interpretable within that framework, with clear policy implications. Federal and state environmental protection and natural resource management agencies in the United States can use it to identify challenges and opportunities for improvement in the institutions they manage and to help make choices when institutional change is possible or is demanded. Participants in collaborative environmental governance institutions can use the research for similar purposes. International lending institutions can use it when they consider which institutional regimes are appropriate for what kinds of environmental problems, at what scales, and in which contexts. Participants in crafting international environmental agreements can use it to gain insight into ways to judge the adequacy of national-level policies for meeting international commitments. All these actors can use comparative institutional analyses to help identify the nature of both problems and assets of existing systems.

Research on governing the commons can sometimes be used in a program evaluation mode, for example when it focuses on the functioning of institutions under the purview of a specific governmental unit (e.g., a forest management plan under the U.S. Forest Service). Its primary value, however, is to provide insights about commons management in general (Ostrom, 1990; National Research Council, 2002b) or with specific resources in specific contexts (e.g., irrigation systems in South India—Wade, 1994; tradable emissions permits for air pollution—Tietenberg, 2002). Such generic knowledge must be interpreted for its implications for decisions at hand. Thus, the main function of this research for decision makers will be to enlighten their choices. It can help identify the approaches that are most promising for a given situation, the governance challenges that are likely to be most difficult, and ways those challenges have been met successfully in similar situations. This research field provides new frameworks for policy analysis that have helped identify key governance problems (for example, the tragedy of the commons) and expanded the set of possible solutions (e.g., refining conceptions of property rights; identifying the potential of hybrid institutional forms). Future research, more contextualized and with greater attention to issues of scale and heterogeneity and the roles of nongovernmental institutions, will enhance and fill in the details of policy analysis based on those frameworks; it may also generate new frameworks or paradigms.

Likelihood of Use

Individual research projects are seldom translated directly into policy choices. However, an accumulation of findings, including challenges to existing assumptions and interpretations of fact, can make a difference in policy (Lindblom and Cohen, 1979). An accumulation of research findings on environmental governance has demonstrated the limitations of policies that indiscriminately promote particular property rules, such as nationalization or privatization: each property regime can lead to either success or failure, depending on how it meets governance requirements (e.g., Feeny et al., 1990, Dietz et al., 2003 and online supplement). Such results appear to have already had some influence on federal marine fisheries policy, in which market-based management regimes are increasingly modified to reflect concerns about both conservation and community (McCay, 2004), and with appropriate attention from decision makers they may become influential in other areas of environmental policy as well. The relevance is clearly there.

The use of research is often influenced by uncontrollable factors, such as its compatibility with the agendas of specific policy leaders and the occurrence of dramatic environmental events that lead to serious reassessments of policies and institutions. Research can nevertheless be organized in ways that increase the likelihood that its results will be used. One way is to encourage a body of research that covers a large enough scale and that continues over a long enough time to become integrated into cumulative changes in understanding and incremental changes in policy (Lindblom, 1959; Lindblom and Cohen, 1979). A recent example with considerable promise is the large comparative case study research project on forest dynamics and governance institutions carried out by the Center for Institutions, Population, and Environmental Conservation at Indiana University (Gibson et al., 2000). Research can also be brought to the attention of policy decision makers by encouraging networks that link the producers and consumers of the research, through formal organizations, such as the International Association for the Study of Common Property (www.iascp.org) and through the participation of researchers in scientific advisory panels and other units of governmental and nongovernmental institutions.

4

The Environment in Business Decision Making

Federal agencies should substantially expand support for research to understand the influence of environmental considerations in business decisions. This research agenda would include studies particularly on (a) environmental performance and competitive advantage; (b) customer and investor demand for environmental performance by businesses, especially in an increasingly global economic system; (c) supply chains and production networks; (d) sectoral standard-setting; (e) decision factors in industrial ecology; (f) environmental accounting and disclosure practices; and (g) government policy influences on business decision making.

THE RESEARCH NEED

Both in the United States and worldwide, business decisions are among the dominant influences shaping environmental conditions: what materials, energy, and organisms will be extracted from the environment, in what quantities and where, how they will be transported and distributed, how landscapes and ecosystems will be transformed in doing so, and what will be done to minimize and mitigate the impacts. Business decisions also influence consumer choices, direct a large fraction of environmental research, and determine much of the development and diffusion of technological innovations.

The cumulative effect of businesses' decisions creates many commitments that are difficult if not impossible to reverse in the short term. Consumer preferences influence these decisions in some cases, but only to the extent that they strongly affect the ability to generate profit. Government policies also influence business decisions—through regulatory mandates, property rights and liability rules, disclosure mandates, taxes and subsidies, procurement criteria, and other policies—but the primary initiative lies with businesses themselves.

To date, however, the role of environmental considerations in business decision making has been seriously understudied. The dominant emphasis of environmental research in the past has been in the natural and health sciences and engineering, addressing such issues as the health risks of particular substances, the functioning of environmental processes and ecosystems and the impacts of changes in them, and technologies for pollution control.

Environmental research in the social sciences to date has concentrated primarily on economics, including the measurement of economic costs and benefits of pollution control to society and the relative efficiency of regulatory mandates versus market-oriented instruments of environmental policy (Stavins, 2003); on government decision making; and to a lesser extent on individual and household environmental decision making, such as energy conservation, recycling, and environmental considerations in consumer behavior (Gardner and Stern, 2002).

Over the past decade, a modest but growing body of research has begun to address environmental considerations in business decision-making (see Appendix C for a review). This research has consisted mainly of a group of literatures associated with established business research fields: environmental considerations in strategic management decisions, in operations, in organizational behavior, in marketing, in accounting, in finance, and in government policies affecting business. Two arguably new research areas also have emerged: one is the study of life-cycle analysis and industrial ecology (the study of flows of energy and materials through systems of industrial production, consumption, and waste disposal), although to date this area has been influenced more by engineers than by the social sciences. The other is the study of supply or commodity chains. Each of these areas, as well as several others, offers promising opportunities for further research.

AREAS OF RESEARCH

The following discussion highlights particularly promising research areas and questions.

Environmental Performance and Competitive Advantage

When does it pay to be green? A central question concerns the conditions under which business decisions that enhance the environment also enhance competitive advantage and other business goals. The answers to this question are fundamental to environmental decisions both by businesses themselves and by governments choosing between regulatory and more market-oriented incentives.

In the past, many businesses considered environmental performance improvement at worst a deadweight cost driven by regulatory mandates and liability risks, or at best an opportunity for cost minimization through more efficient use of materials and energy (Royston, 1979, 1980; Andrews, 1999). More recent studies have begun to document that competitive advantages can be gained through environmental protection under at least some conditions (Hart and Ahuja, 1996; Klassen and McLaughlin, 1996; Klassen and Whybark, 1999). Many of the most promising environmental improvements in business performance may result not from incremental changes at existing factories but from investments in new facilities ("ecological modernization") and in new products and technological innovations (Hart, 1997; Hart and Milstein, 1999). New factories typically are far more efficient than older ones, including greater efficiency in the use of materials and energy and in pollution prevention and waste reduction, and many new products also include less use of toxic materials, as well as less energy and materials than older ones, although the aggregate human consumption of materials and energy nonetheless continues to rise. Van Heel, Elkington, Fennell, and Franceska (2001) identify 10 distinct dimensions of business value against which environmental performance can be measured: shareholder value, revenue, operational efficiency, access to capital, customer attraction, brand value and reputation, human and intellectual capital, risk profile, innovation, and license to operate.

When environmental protection confers market advantage, does such advantage derive chiefly from external incentives, such as customer or investor demand, governmental requirements and subsidies, or social and community pressures? Or does it derive also from distinctive capabilities and resources of the firm itself, as a growing body of business research now suggests (Hart, 1995; Russo and Fouts, 1997; Prakash, 2001; Aragón-Correa and Sharma, 2003)? Further research is needed to characterize more precisely why some facilities and firms create significantly greater competitive advantage through superior environmental performance than others even in the same sector and to identify how internal capabilities as well as external pressures influence those outcomes.

How do competitive advantages from environmental protection, where they exist for individual firms, influence environmental protection more generally? Do these practices gradually disseminate to other firms, reducing the initial competitive advantage of first movers but improving overall environmental outcomes? Do the high-performing firms use their competitive advantage to gradually displace poorer-performing competitors, following Schumpeter's theory of creative destruction (Hart and Milstein, 1999)? Or do high-performing firms gain their competitive advantage simply in high-end niche markets, while poorer performers continue to coexist with them in other markets and with little overall improvement in environmental outcomes? To understand the overall environmental effects of environmentally protective actions by firms, it is essential that we understand not only the behavior of the most environmentally innovative and competitive firms, but also their impact—and the limits of that impact—on the environmental performance of other firms (Sharma, 2002).

Why are many businesses less receptive to environmental initiatives than the evidence for business benefits suggests they should be, and under what conditions do such patterns change? Even when green practices such as waste reduction and energy conservation have demonstrated positive rates of return, businesses often have been slow to adopt them. Organizational inertia is one possible answer (Sarokin, Muir, Miller, and Sperber, 1986): environmental responsibilities often are isolated in an environment, health, and safety unit responsible only for regulatory compliance and with little influence on broader management and investment decisions. In this regard, a potential benefit of the increasing use of the ISO 14001 environmental management system standard has been its influence in mainstreaming environmental considerations into the responsibilities of all senior managers and thus increasing integration of such considerations into core business decisions. In some instances, however, when business benefits exist for green initiatives, these benefits may be smaller than those of other investment opportunities or managerial priorities (see, e.g., Feder, 1999; Greer, 2000). Better understanding of such issues may shed light on more fundamental patterns of imperfect rationality and suboptimal behavior in business organizations, in addition to their benefits for understanding businesses' environmental performance.

Customer and Investor Demand

What factors affect customer and investor demand for environmental performance? Research on this question would focus on the circumstances under which influential customers and investors care about environmental performance, how such preferences influence or fail to influence business decisions that more directly affect environmental outcomes, and what brings about change in the influence of these demand factors.

Customer and investor demand are key drivers of business decisions. A research literature has developed on consumer demand for green products at the level of individual choice, including related topics such as the effectiveness and value of green labeling and product certification (e.g., Thøgersen, 2002; see also Chapter 5). A modest literature also has begun to emerge on environmental performance and capital markets (see http://www.institutionalshareowner.com/research.html).

How is the environmental performance of businesses likely to be affected by future demand for environmental performance in international and emerging markets? Even for U.S. businesses, future products and production processes are increasingly driven by global rather than merely domestic markets. Much of the modest literature so far has focused on U.S. and European markets, and even here important questions remain unanswered, such as the reasons for the apparently greater demand for environmental performance among European than U.S. consumers, the impacts of these norms on U.S. firms as well as others, and whether these patterns will continue as the European Union itself expands to less affluent and more diverse countries. Beyond the European Union, the major markets for the future will be in large emerging and industrializing economies, such as China. How will their consumers' and investors' demand influence environmental performance, and how will these demands both influence business decisions and change with economic growth?

How will businesses be affected by the demand for better environmental performance by influential business partners, not only by household consumers, as implied above, but also by business customers, institutional procurement offices, partners in joint ventures, suppliers, insurers, financiers and large institutional investment funds, and others? Business partners may exercise different preferences than end-use consumers, driven by their own calculations of efficiency, profit, liability, reputational risk, and other business considerations. Home builders, for example, may order appliances that are low in capital cost but high in energy operating cost to the housing consumer in part because few consumers consider the energy costs of operating home appliances when they are buying a new house. Some business partners have begun to be more demanding than this consumer practice suggests, requiring explicit evidence of environmental performance as a condition of doing business, for example, through environmental risk reviews by lenders and insurers, environmental management commitments by some major corporations, and social screening criteria used by some investment and pension funds. Other business partners continue to reward price over environmental performance.

In short, for some businesses, demand by major customers or other business stakeholders may be more influential than end-use consumers' demand. A fruitful line of further research would therefore be to explore the circumstances under which major business partners have this influence and have exercised it in favor of environmentally protective preferences, and the extent to which such influence could affect other environmentally significant products and production processes of their suppliers.

Supply Chains and Production Networks

Whose business decisions drive environmental outcomes? Research on this question concerns environmental decision making in supply or commodity chains, representing sequences of production decisions across multiple businesses from raw materials extraction to retail sales, consumption, and post-consumer waste management.

Many of the greatest environmental impacts are driven not by the decisions of individual factories or even firms, but by decision rules embedded in larger scale chains or networks of production relationships. These chains involve significant externalities, in which the firms that cause the most serious environmental impacts—for example raw materials extraction and processing—often are not directly identifiable by consumers or investors who are interacting directly with another company in the chain. Moreover, the choices of the firms that have the most serious direct environmental impacts may be seriously constrained by the decisions of dominant firms at other points in the value chain, such as major retailers.

Past research has focused primarily on the individual firm or industrial sector as the unit of analysis, but not on the larger institutional networks of decisions, incentives, and constraints. Recent research has begun to characterize influence relationships in supply chains with greater theoretical precision (Gereffi, Humphrey, and Sturgeon, 2002; Humphrey and Schmitz, 2001) and to examine their implications for environmental performance (see Appendix C for more detail).

Under what conditions do the demands of dominant businesses create incentives for, or barriers to, environmental improvement by other participants in commodity chains? For example, firms operating in countries with high environmental standards may be tempted to externalize environmental impacts to other firms through relocation or contracted outsourcing (Bommer, 1999). Alternatively, such firms might pressure their suppliers for high performance as well, for reasons including efficiency in logistics, minimizing potential liabilities, or simply enhancing their public image and brand value (Vogel, 1995; Garcia-Johnson, 2000). Under what conditions do certain firms in supply chains act as weak links with respect to environmental improvement, limiting efforts of other firms in the chain to improve the environmental performance of the entire productive process?

Supply chain mandates have recently emerged as a new mechanism for leveraging environmental performance improvement (Andrews, Hutson, and Edwards, 2004). The effectiveness and generalizability of such mandates, however, has not yet been well documented. Will businesses actively enforce them? How high a standard will they require? Will their effects extend beyond direct suppliers all the way to raw materials producers, the point at which many of the most severe impacts occur? These questions raise complex issues and further research questions involving the difficulty of labeling and tracking raw materials as commodities, the multiplicity and diversity of suppliers, and open-market purchasing of some product components. In short, supply chain research may shed new light on important aspects of business decision making for the environment, but it may also reveal important limits to efforts to use supply chain leverage as an alternative to direct protection of at-risk environments and ecosystems.

Considerable insight may be gained from research that links issues of supply chain governance with the established research field of product life-cycle analysis (Franklin Associates, 1991). Life-cycle analysis has developed as a technical field of environmental research, evaluating the environmental consequences of a product from "cradle to grave" by identifying the impacts of each of its component processes (see http://www.umich.edu/~nppcpub/resources/compendia/CORPpdfs/CORPlca.pdf), but often without concurrent research on the supply chain decision making that determines each of these choices. Supply chain research has now advanced to the point at which promising connections could be made between these fields.

Important insights are also likely to come from research on the downstream ends of supply chains, such as on the substitutability of services for some products, the logistics of returning recycled products into production processes, and the implications of emerging public policy mandates for post-consumer recycling of major classes of products (for example, the European Union's recent mandates for product stewardship and for recycling of waste electronic products).

Sectoral Standard Setting

Can trade associations effectively serve as nongovernmental regulators of environmental decision making by their members? In some industrial sectors, such as chemicals and forest products, a dominant business or trade association has incorporated environmental codes of conduct or performance and certification standards into membership requirements (Nash, 2002). Such initiatives represent attempts by business communities at self-policing, perhaps to forestall government regulation but also to protect a shared reputation or "club good" (Kollman and Prakash, 2002; Potoski and Prakash, 2005). In some other sectors, trade associations have functioned as important agents of environmental performance improvement in waste reduction, pollution prevention, and environmental management more generally (for example, the Environmental Protection Agency's Sector Strategies Program, http://www.epa.gov/fedrgstr/EPA-GENERAL/2003/May/Day-01/g10887.htm).

How strong do such sectoral codes and commitments prove to be, and how effectively enforced, balancing competitive advantage for the best firms against collective-action temptations toward the lowest common denominator? Do they affect member firms' expectations toward their suppliers and customers? And in what other sectors might such club-good characteristics be exploited to produce better environmental performance?

What are the major barriers to improved environmental performance in particular understudied but environmentally significant sectors? Buildings, for example, once constructed, are among the most significant long-term drivers of energy consumption and its environmental impacts. There is a growing body of research on cost-effective technologies for reducing those impacts (Kats, 2003), but both private and public funding for building-related research lags behind other major sectors, and green-building technologies so far have penetrated only modestly into the construction and development industries (U.S. Green Building Council, 2003).

What has not yet been well studied is the role of key institutional barriers to environmental performance improvement in this sector, such as the decentralized structure of the sector, fragmented markets and regulatory jurisdictions, cost-analysis practices dominated by initial construction rather than life-cycle costs, and effective steps for reducing these barriers. Sector-specific barriers may also exist in other environmentally significant sectors, such as many service sectors.

Decision Factors in Industrial Ecology

What decision factors are critical to the redesign and optimization of industrial processes for environmental benefits? Industrial ecology has recently emerged as a new domain of research embracing the multidisciplinary study of industrial, technological, and economic systems and their linkages with fundamental natural systems (Socolow, Andrews, Berkhout, and Thomas, 1994; Graedel and Allenby, 2003). Foreshadowed by the work of Leontief (1966) on input-output analysis, industrial ecology seeks to understand, and in its normative application to optimize, the environmental and economic outcomes of entire industrial processes (Socolow, 1994). Frosch and Gallapoulos (1989) suggested that industrial systems could be more efficient if their material flows were modeled after natural ecosystems; since that time, the research literature on this idea and its potential applications has proliferated (Ayres and Ayres, 2002; see also the *Journal of Industrial Ecology*, established 1997). A recent National Research Council report concluded that analyses using material flows data have already proven useful, and that a more systematic set of material flow accounts across the economy would benefit improved public policy, the efficiency of business decision making, environmental and economic performance, and national security (National Research Council, 2004c).

Much of the research on industrial ecology to date has focused on technological and engineering factors in particular manufacturing sectors (Allenby, 1992, 1999; Graedel and Allenby, 2003; U.S. Congress Office of Technology Assessment, 1992). As yet there has been far less research into the business decision-making processes and other factors affecting the implementation and wider potential of applied industrial ecology.

A promising topic for further research would be to examine systematically the instances in which industrial ecology innovations are well established and to identify the decision factors, social processes, and other circumstances that were essential to successful implementation. From the findings, researchers could develop a theoretical model for predicting other sectors and circumstances in which such successes could be successfully introduced. There is a substantial increase in analytical complexity, however, when decision making moves from the regime of traditional pollution-control technologies, such as air scrubbers and water treatment plants, to core production technologies. The most obvious difference is the degree to which the technologies involved are coupled to other technologies (Shapiro and Varian, 1999).

In the case of an air scrubber, for example, one can generally change technologies without affecting any of the underlying manufacturing technologies, which are only loosely coupled. But eliminating a major process material—lead solder in electronics manufacturing, for example—has far more complex implications, since the technology associated with such a material is more tightly coupled to other technologies in the design and manufacturing process and significantly affects product performance. Nonlead solders may require more stringent cleaning of the underlying copper substrate, for example, which could necessitate changing from water-based to chlorinated-solvent cleaning systems; the latter would have potential health impacts that aqueous cleaning systems do not, posing air quality issues as well. Moreover, the bond that is formed by the new joining technology might not be as robust, requiring changes in physical design and in the use environment. And at the end of the product's life, care must be taken that the new formulation does not degrade recycling economics or technologies. Some metals "poison" copper, for example, so that it cannot be reused in electronics, and such metals must be avoided in solder formulation (Allenby, 1992).

The reality of such couplings among technologies creates a form of technological lock-in. Once a new technology is integrated into the design, manufacturing, use and disposal cycle, a subsequent change will be much more difficult to implement, for further change in any single technology may affect others as well. Core technologies thus evolve at different rates, reflecting the complexities of material substitution at industrial scale (National Research Council, 1999f). When the electronics industry was challenged to shift from chlorofluorocarbons and chlorinated solvents to aqueous cleaning systems in components manufacture, it was able to do so in a few years. But the shift away from lead solder has already taken a decade and a half and is not yet complete. Similarly, one can readily build a hydrogen-based car today, but the technology system required to get the hydrogen to it on a mass market basis doesn't exist: the gasoline engine is locked in by the fuel manufacture and distribution infrastructure. Change therefore takes much longer, is more complicated and difficult to achieve, and has much more potential for unanticipated consequences as it ripples through the coupled technological, economic, and social systems (Bijker, Hughes, and Pinch, 1987).

These observations suggest two promising directions for social science research. First, there would be significant value in research that provides environmental decision makers with a greater understanding of technological systems and their evolution. Second, research that studies the technological context in which environmental decisions are made, including the reflexive relationship between environmental requirements and subsequent technological evolution, could be valuable not only for making environmental decisions more informed and effective, but also for anticipating whether particular choices will either create net environmental value or generate systems effects that reduce or even outweigh environmental benefits.

Environmental Accounting and Disclosure Practices

How can environmental performance best be measured and reported for use in business decision making? It is a truism among business executives that "what gets measured, gets managed." Environmental issues pose important challenges to traditional accounting and management information practices. Some environmental impacts of a firm's activities represent significant hidden costs and unrecognized opportunities for economic benefit (for example, regulatory compliance and waste management costs, worker's compensation and other insurance liabilities). Others represent significant but often unrecognized hidden liabilities, such as contaminated sites or toxic substances in products, as well as unrecognized opportunities for economic gain.

Research to identify such costs and business opportunities more explicitly, and to attribute them more specifically to the processes that generate them, would represent logical extensions of recent innovations in management accounting, such as activity-based costing and full-cost accounting. There are significant unresolved issues associated with such changes, however, that also require research. Appropriate measures must be developed for estimating and charging such costs, and standards developed regarding which environmental costs should be formally recognized and disclosed to investors and the public, and at what point, in order to provide an accurate and appropriate picture of the company's assets, liabilities, and performance (Kirschner, 1994). Because these research areas focus on management and financial accounting, they are distinct from other research that focuses on counting environmental resources as capital assets in national income accounts (Ahmad, El Serafy, and Lutz, 1989; National Research Council, 1999c).

Implicit in any effort to improve environmental accounting practices is the need to develop credible measures of the impacts of environmental performance on economic performance. There is a modest but growing body of empirical work on this subject (e.g., Dowell, Hart, and Yeung, 2000; Hart and Ahuja, 1996; Klassen and McLaughlin, 1996; Klassen and Whybark, 1999; Levy, 1995; Russo and Fouts, 1997; van Heel et al., 2001). Early results suggest positive linkages among environmental, social, and economic performance outcomes, but the numbers of studies remain relatively small and the research is not yet systematic in its coverage of entire sectors or economies and relatively short-term in its time span. More systematic investigation of such measures would be useful, particularly in sectors subject to significant environmental impact and high variation in environmental performance among firms (see Appendix C for further discussion).

A related research need concerns the standardization of environmental disclosure practices. Environmental performance information to date has been reported primarily in voluntarily prepared environmental and sustainability annual reports, and it is not consistent or comparable across firms. Recent initiatives have sought to promote standardization and greater quality control of this information (Global Reporting Initiative, www.globalreporting.org), on the ground that such standardization would be in the interest not only of the public but also of firms (a single reporting format would be more efficient for all firms and would offer competitive advantage and benchmarking opportunities for superior performers).

There remain important unresolved questions for research regarding the design of reporting standards. Key questions include the appropriate measures and range of indicators to be generally used and approaches to aggregating them into overall indicators of environmental performance. As practical matters, both the verifiability of such measures and the potential liabilities associated with underreporting, overreporting, or inaccurate reporting also remain unresolved concerns.

Government Policy Influences on Business Decision Making

What are the net overall incentive effects of government policies on business decisions affecting the environment? Substantial research literatures already exist on the direct effects of regulations and economic incentives, such as environmental taxes, charges, and tradable permits (Stavins, 2003), as well as the direct effects of information disclosure requirements (Tietenberg, 1998). There also is a growing literature on the effects of government-supported voluntary initiatives, in which the government offers technical and informational assistance, certification standards, favorable public recognition, and/or increased regulatory flexibility in exchange for commitments by the firm to reduce its environmental impacts beyond what is required by regulation (Andrews, 1998; Mazurek, 2002; Harrison, 2002).

Far less research has been conducted, however, on the enhancing or offsetting effects of other government policies that also affect business incentives. For example, in the United States both regulations and tax advantages favor end-of-pipe pollution-control technologies over innovation in production processes, whereas technology policies often lack explicit environmental criteria. A recent workshop report by the Organisation for Economic Cooperation and Development argued that decisions about technological innovation lie at the heart of business decision making about improving environmental performance and concluded that one of the most important public policy challenges is to coordinate the technological incentives of environmental policy with the environmental effects of innovation policy (Organisation for Economic Co-operation and Development, 2000:22, 14). Norberg-Bohm (2000), after reviewing the literature on environmental policy influences on green innovation by businesses, concluded that research on the combined effects of multiple policy mechanisms on business decision making would be particularly useful.

When policy incentives are used to promote green innovation, under what conditions do they have the undesired effect of locking in and privileging particular technologies? Both government regulatory and procurement standards and standards set by business organizations are vulnerable to strategic behavior by businesses to lock in mandates that favor their own interests and products (Pashigian, 1985). The modest amount of research and the more extensive anecdotal experience on this subject needs to be consolidated into empirically based principles by which performance-based policy instruments intended to favor better environmental outcomes also create incentives for open competition in further innovation.

What are the global effects of shifting government standards on business decision making? In the 1970s and 1980s, U.S. standards were the dominant approach adopted (whether or not enforced) by many countries and assumed by most transnational businesses worldwide. Increasingly, however, the European Union has moved into this norm-setting role (Vogel, 2003), and China is now emerging as an important enough production location and potential consumer market that its standards are also likely to play a major role.

As an example, the United States and the European Union have developed different systems for reviewing the risks from new chemicals. The European Union has specific testing requirements, but the requirements are imposed only after the volume of production of the chemical reaches a specified level. The United States has no specific testing requirements, but the requirement to submit any available data on the chemical is imposed prior to the chemical being manufactured. This difference may put a premium on trying out a new chemical in Europe. However, a proposed European Union regulation, the REACH initiative, would put the burden of proof of safety on chemical manufacturers and could substantially change the calculation for manufacturers. Other examples involve differing norms for content and recycling of electronic products and for introducing genetically modified food crops.

Changes in the leaders in standard setting may create important shifts in business expectations, in competitive advantage among businesses operating in these markets, and in worldwide norms for environmental decision making by businesses. They may also produce a gradual convergence of regulatory expectations, but little is yet known about whether the likely convergence would be either environmentally beneficial or economically efficient. More systematic research could shed light on these important issues.

RATIONALE FOR THE SCIENCE PRIORITY

Likelihood of Scientific Advances

Over the past several decades, the implications of environmental considerations for business decision making have become more widely recognized by leading business decision makers than by the research community. Yet because of the paucity of systematic research, decisions by these business leaders have been based largely on anecdotes and driven by a focus on limited elements of these implications—regulatory constraints, cost burdens, and liability risks, for example—rather than by analysis of the full range of opportunities for adaptation and innovation that could benefit businesses as well as the public.

Significant scientific advances can now be anticipated in this priority area for several reasons. First, research over the past decade has begun to produce solid theoretical, methodological, and empirical foundations for such investigations, beginning with a research agenda posed by the Greening of Industry Network over a decade ago and continuing with the emergence of a productive and growing community of scholars since (Fischer and Schott, 1993; Sharma, 2002; also see Appendix C). Many of the topics proposed would build directly on theoretical foundations that have now become available, such as on supply chain governance, industrial ecology, and the resource-based theory of the firm.

Second, the topics proposed offer opportunities to link and integrate bodies of research that have so far developed separately, such as supply chain governance and life-cycle analysis, and to extend research that has developed in a U.S. domestic context to the increasingly global economic and institutional context in which many environmental outcomes now are shaped. These linkages are likely to lead to new insights.

Third, many businesses themselves have become more actively interested in these questions and more willing to share with researchers the information necessary to conduct empirical research on them. Examples include large-scale surveys, detailed case studies, and some studies involving extensive access to internal documents as well as publicly reported data (e.g. Andrews et al., 2001).

Potential Value

Research on environmental considerations in business decision making is of direct interest to several user groups. First, it is of interest to many businesses, particularly leading firms in economic sectors that have significant environmental impacts, as well as other aspiring firms seeking to benchmark themselves against their competitors and against "best-in-class" firms in their industries (see, for example, the web site of the World Business Council for Sustainable Development, a voluntary association of self-identified leading firms in environmental and social performance, www.wbcsd.org).

Second, such research is of interest to many sectoral trade associations, business research organizations, and others that serve subgroups of the business community. Several trade associations, for example, have mandated environmental codes of conduct as membership criteria (e.g., the American Chemistry Council and the American Forest and Paper Association); others have sought to assist small enterprises in their sectors to improve their environmental performance (metal plating, dry cleaning, and automotive repair shops, for example). Such interest is particularly evident in sectors in which environmental performance is less a matter of competitive advantage and more an issue of widely shared costs, liability, or potential reputational damage to the industry as a whole (Kollman and Prakash, 2002). These sectors and associations have an interest in the credibility of their environmental codes and in inducing compliance with them.

Third, such research is of direct interest and value to government agencies, particularly to many that have begun to promote voluntary approaches to environmental performance improvement by regulated businesses. Essential to such initiatives—and indeed to more effective regulation and other public policy incentives for environmental performance improvement—is obtaining a more accurate understanding of the factors actually influencing business decision making, and of their variation across firms and sectors, so that public policy incentives can be more accurately, reliably, and efficiently designed to achieve their intended effects in improving environmental quality.

Fourth, such research is also important for informing the interested public about which environmental performance claims by businesses can be relied on and about which businesses are in fact improving their performance as opposed to merely burnishing their images or outsourcing damaging activities to less visible suppliers.

This science priority provides a valuable agenda for government research agencies. In recent years, the Environmental Protection Agency's National Center for Environmental Research has begun to support research on a number of questions related to business decision-making affecting the environment under its STAR (Science To Achieve Results) Economics and Decision Sciences program (see http://es.epa.gov/ncer/science/economics/). It has specifically noted as priorities research on motivations for corporate environmental behavior and on behavioral responses by businesses to government interventions (such as regulatory compliance and implementation of voluntary programs), among other areas. Most research so far under this program has been limited to regulatory compliance behavior, environmental performance measurement, and performance changes associated with voluntary initiatives, such as environmental management systems. This science priority would expand the research under this or related programs to address other promising topics as well.

The National Science Foundation also has demonstrated interest in research on business decisions affecting the environment. Its program for decision, risk, and management science supports research in management science, risk analysis, societal and public policy decision making, behavioral decision making and judgment, organizational design, and decision making under uncertainty, including particularly work on judgment and decision processes; risk perception, communication, and management; organizational performance; and modeling of managerial processes. The NSF program on Human Dimensions of Global Change includes explicit research priorities on innovation and diffusion processes related to global environmental change, resource use and management, anticipatory and reactive adaptation and mitigation, economic issues including international trade patterns and global sectoral models, and environmental accounting. All these areas would allow for research on environmental considerations in business decision making. This science priority would expand the focus on business decision making in the above research programs. Research on these questions would also be valuable in other countries, as indicated by interest in these questions among agencies in the United Kingdom and European Union (see http://remas.ewindows.eu.org/index.htm/REMAS/reports/relatedreports.htm#brit).

Likelihood of Use

The research proposed here could produce not only advances in scholarly understanding but also practical improvements in the environmental performance and in the efficiency and competitiveness of businesses. Poor environmental performance often represents economic waste and other forms of competitive disadvantage as well, particularly in the face of rising community expectations and regulatory demands in major markets, such as the European Union (Vogel, 2003).

It is true than many firms are far more preoccupied with immediate profitability than with environmental considerations and that many are unlikely to pay attention to research, except perhaps market research related to their own products. But these generalizations do not apply to many of the leading firms in key industries, which are keenly interested in ways to improve their environmental performance. Many of the leading firms are also major transnational corporations that can promote wide dissemination of research findings through their influence on their subsidiaries, suppliers, and business customers, on their peers in the same and other sectors, and even on government decision makers and the news media.

There is good evidence that leading companies are actively interested in information related to environmental performance. A recent survey by the Conference Board, for example, found that more than half the responding firms explicitly reported environmental performance at least annually to their board of directors; well over half tied their compensation of plant and operations managers to environmental performance, and more than 40 percent did the same for executives and senior managers. Nearly half also had a mechanism in place to drive design processes toward minimizing overall environmental impacts in product or process development (Lowy and Wells, 2000). More than 40 percent of these firms believed that effective environmental governance added value to their operational efficiency as well as their corporate image, and 28 percent reported increased value to their competitive position.

The Conference Board survey also found that companies identified as leaders by their peers in other firms were 2.5 times more likely than other firms to leverage their organization's core business competencies to address strategic environmental concerns; those identified leader organizations were 1.5 times more likely to consider their environmental expertise itself a core competency in formulating their overall business strategy than were the other respondents. Nearly 90 percent of peer-identified leaders (compared with 38 percent of other respondents) had mechanisms in place to drive design processes toward minimizing overall environmental impacts during the product or process concept development stage while addressing customer needs. Two-thirds of the leaders (compared with 37 percent of the others) subjected all product designs to a final environmental review and approval prior to market introduction. And leaders were more likely to have mechanisms in place to minimize life-cycle impacts of specific product or process designs through such techniques as design for environment or life-cycle analysis.

Leading firms tended to see environmental challenges as opportunities for product development, whereas other respondents tended to see them as regulatory threats. Leading firms also far more consistently reported increases in value to their corporate image (86 percent), access to their communities (57 percent), public opinion (43 percent), and business opportunities (43 percent) (Lowy and Wells, 2000).

The findings of this survey suggest strongly that at least leading firms would have a strong interest in the results of research such as that proposed here, and that such firms also are recognized as benchmarks and potential role models by others. Having such research supported by government agencies would increase the credibility of the results. Moreover, for many of the research questions raised here, studies sponsored by private-sector organizations, if conducted at all, may not address issues of public interest as effectively as government-supported research.

The recommended research may also be used by government decision makers, many of whom have shown interest in improving their understanding of business decision making in order to design more effective and efficient programs for environmental performance improvement. Examples include the recent proliferation of government-sponsored voluntary initiatives (Mazurek, 2002) at both the state and federal levels, as well as increased use of market-oriented and disclosure-based incentives as policy tools intended to promote improved environmental performance by businesses (Stavins, 2003; Tietenberg, 1998).

Finally, an additional reason for government use of this research is the increasing pressures for environmental performance improvement—and improved management and decision making more generally—on the part of many government-operated business enterprises themselves, such as water and wastewater treatment utilities, hospitals, military facilities, and others. For example, presidential Executive Order 13148, issued April 22, 2000, requires formal environmental management systems for all appropriate federal facilities.

5

Environmentally Significant Individual Behavior

Federal agencies should support a concerted research effort to better understand and inform environmentally significant decisions by individuals. Research in four specific areas could provide usable results in the relatively near term: (1) indicators of environmentally significant consumption, (2) information transmission systems, (3) integration of information with regulatory and market-based policy instruments, and (4) fundamental understanding of consumer choice and constraint. Research in the first and the last of these areas is likely to have the most important and lasting impact.

THE RESEARCH NEED

The activities of individuals have major environmental consequences in the aggregate. Consequently, there can be major environmental effects from change in the behavior of individuals and households. The often-cited estimate that consumer expenditures account for two-thirds of gross domestic product suggests the environmental importance of consumer power, although it is likely to be an overestimate of the direct environmental effect of consumer behavior. Households directly account for slightly under half of U.S. carbon emissions and smaller percentages of some other important effluents (e.g., Stern and Gardner, 1981; Cutter et al., 2002). For several pollutants that have been greatly reduced by effective regulation of the industrial sector, individuals and households have become major sources of the remaining emissions (Vandenburgh, 2004). Dioxins and dioxin-like compounds are a prime example. Between 1987 and 2002, total emissions of these compounds in the United States were reduced by 92 percent, with the result that backyard barrel burning, which was the source of 4 percent of emissions in 1987, had become the source of almost 60 percent of national emissions 15 years later (Institute of Medicine, 2003).

Individual behaviors have significant direct impact in the aggregate in the areas of transportation, housing, energy-using appliances, solid waste, water, and food. Individuals also influence environmental quality indirectly, in their roles as citizens, investors, and members of organizations that make environmentally important choices. And to an important degree, small businesses and nonprofit organizations have impacts (and make decisions) much like individuals and households. However, these latter areas are not central to the scope of the panel's consideration of individual environmental choices.

The environmental impact of an aggregation of individual consumer choices is circumscribed because, in many cases, the links from individual behavior to its environmental consequences are indirect and conditioned by a variety of forces and constraints in complex social, economic, institutional, and technological systems (Shove, Lutzenhiser, Guy, Hackett, and Wilhite, 1998; Lutzenhiser, Harris, and Olsen, 2001). For example, individuals who want to make green choices may find options limited and costs prohibitive because of a lack of the relevant products or infrastructure, as is generally the case in the United States with alternatives to petroleum-burning private motor vehicles for personal transportation. When green choices are more readily available, the potential for environmental improvement may be limited because of the difficulty or cost of acquiring trustworthy and timely information about the environmental consequences of decisions. This continues to be the case, even though an increasing number of products and services are marketed with claims that they are environmentally benign or beneficial.

The links from policy to individual behavior are also weaker than sometimes supposed. For example, governments sometimes provide information in their efforts to promote greener individual behavior—increased recycling, more careful use of household chemicals, purchase of energy-efficient appliances, testing for radon in homes, reduced use of motor vehicles during air pollution crises, and so forth. They presume that, with better information, people will act in more environmentally beneficial ways. But the record of environmental information programs is unimpressive (Hirst, Berry, and Soderstrom, 1981; National Research Council, 1984, 2002a). Information can be more effective, however. Research has identified some of the ways in which information, usually in interaction with a variety of other factors, influences environmentally significant choices (e.g., National Research Council, 1984, 1997a, 2002a). It shows that environmental information can be effective in influencing behavior if it is delivered in appropriate and timely ways. The ways in which information is constructed, conveyed, cognitively processed, and weighed (in interaction with other factors, such as individuals' values, attitudes, and affective processes; social norms; economic resources and incentives; and technological availability) determine whether, when, and how information affects action.

The actual environmental impact of effectively delivered information on individuals' choices is not frequently measured, but it is in many cases likely to be modest because of the other factors affecting these choices. Improving information for individual choice is nevertheless important. In some arenas, informing citizens' decisions is a basic duty of government. Also, environmental policies and programs that rely on regulatory or incentive strategies can fall far short of their potential if their information components are not implemented effectively. For example, identical incentive programs operated by different energy utility companies differed by more than a factor of 10 in their effects as a function of how the programs were marketed (Stern et al., 1986). We give special attention to information in this science priority because this policy strategy has significant untapped potential, because its perceived appropriateness does not fluctuate sharply with events and political shifts, and because, in some policy arenas, improving information is among the few policy strategies available.

AREAS OF RESEARCH

Despite a thin stream of research over several decades (e.g., National Research Council, 1984, 1997b; Gardner and Stern, 2002), we are only beginning understand the ways individuals' values and preferences, social influences, available choices and constraints, and other factors combine to shape environmentally significant individual behavior. Developing this fundamental understanding and improving measures of the environmental effects of consumer behavior are critical needs for anticipating the aggregate environmental effects of individual behavior and for informing policies affecting this behavior. We have identified four specific research areas as worthy of particular attention.

Indicators of Environmentally Significant Consumption

Research to develop trustworthy indicators of the environmental impacts of individual and household behavior and choice would help people make their environmentally significant choices conform better to their values and preferences. Many people would like to reduce the environmental impact of their personal choices. Evidence of this desire comes from survey research since the 1970s that, despite periodic fluctuations in the numbers, has recorded very high rates of public support for environmental protection consistently since the 1970s (Dunlap and Scarce, 1991; Dunlap, 2002). For example, 44 percent of U.S. respondents surveyed in 2001 said they thought environmental regulations had not gone far enough, compared with 21 percent who said they had gone too far. The comparable figures were 34 and 13 percent in 1973, when the question was first asked. On the issue of whether government is spending too much, too little, or the right amount on environmental protection, respondents favored "too little" over "too much" by 61 to 7 percent in 1973 and 62 to 7 percent in 2000 (Dunlap, 2002). Evidence of individuals' support for environmental protection also comes from the growth of support for and participation in recycling programs and tangible support for, and widespread use of, parks and nature reserves, community gardens, conservation organizations, farmers' markets, "smart growth" policies, environmentally sensitive products and foods, and other green goods and services. Indirect evidence comes from the prevalence of green appeals in advertising, although it is also the case that advertisers often highlight the nonenvironmental attributes of products and services that have large environmental "footprints."

Against this evidence of strong environmental concern in the United States is apparently contradictory evidence from consumer behavior, such as the continuing growth in sales of fuel-consuming sport utility vehicles, in the size of new homes, and in other trends that increase the environmental impact of Americans' behavior per capita and per household. This apparent paradox of green attitudes contradicted by behavior can be illuminated by better measurement and presents important questions in its own right, as noted later in this chapter.

Although many of the consequences of individuals' choices are easy for them to discern, it is very difficult for individuals who want to consider the environmental impacts of their choices to estimate those impacts. Measures of economic consumption do not adequately capture them (National Research Council, 1997b), and the readily available direct measures have proved less than adequate. For example, utility bills report aggregate energy consumption but do not link it to particular appliances or behaviors. It is therefore not surprising that many consumers hold mistaken beliefs about matters as straightforward as the relative energy consumption of household appliances (e.g., Kempton, Harris, Keith, and Weihl, 1985). Thus, it is important to distinguish, both conceptually and methodologically, between consumers' intent with regard to environmentally significant behavior and the environmental impact of the behavior. Energy-use indicators are provided on labels on major household appliances in the United States, but they do not communicate well with consumers (duPont, 2000; Egan, Payne, and Thorne, 2000; Shorey and Eckman, 2000). Green labels for other consumer products have not developed much in the United States, but there is some evidence from Europe that they can influence individual behavior and perhaps also the behavior of firms (Thøgerson, 2002). A striking example is the eco-labeling of household chemicals in Sweden beginning in the late 1980s, which was followed within a decade by a 15 percent decrease in national sales of cleaning and personal care chemicals and in the replacement of 60 percent of the chemicals in soap, shampoo, detergents, and cleaners by less harmful substances. The success of such labeling programs depends, among other things, on their accuracy and on the existence of an effective information delivery system (see below).

Green-minded consumers need information not only to compare different brands or models of the same product, but also to compare different behaviors (e.g., commuting by car versus bus), to identify environmentally superior products (e.g., paper versus plastic bags), and to determine which environmentally motivated behaviors would have the greatest beneficial environmental effect. Conceptually, this information would amount to scaling down indicators such as the ecological footprint (Rees and Wackernagel, 1995) or the Toxics Release Inventory to the individual level.

Trustworthy information on the environmental impacts of consumer choices depends on the availability and continual improvement of appropriate impact measures, such as direct energy consumption, embedded energy, material flow measures, and life-cycle environmental impact of both products and activities. The recommended research would make such measures behaviorally relevant by linking them to important consumer choices. Such indicators are not now readily available, although increasingly there are efforts to develop them, building on the work of researchers and environmental groups (e.g., Uusitalo, 1986; Durning, 1992; Vringer and Blok, 1995; Lutzenhiser, 1997; Brower and Leon, 1999; Redefining Progress, 2004). The need is to develop, validate, and gain widespread use of sound, behaviorally relevant measures.

Careful accounting studies of materials and energy flows that combine physical science expertise and an understanding of human behavior patterns can provide useful information about the relative impacts of common behaviors and choices, as well as significant alternatives (driving vehicles with various attributes, using private vehicles versus public transit, adoption of different home heating and cooling technologies and retrofits, eating different sorts of foods, eating at home versus in restaurants, using different sorts of appliances, buying durable versus disposable consumer products of various kinds, buying bulk versus prepackaged products, recycling, conserving water, etc.). It may be possible to aggregate the environmental indicators for various consumer products and services into indicators for the companies that provide these goods and services in ways that could usefully inform individuals' investment decisions.

A good accounting of the environmental impacts of behaviors, products, and so forth can better inform individuals' environmentally significant choices. It is likely to reveal that certain practices that are widely believed to be pro-environmental may turn out to have fewer benefits than other, less obvious, choices that may be much more environmentally significant. As a general rule, for example, the environmental impact of purchases of consumer durables such as motor vehicles and major appliances that lock in environmental consequences for long periods is considerably larger than that of variation in how these items are used (Gardner and Stern, 2002). Efforts to develop indicators that are applicable internationally will also be useful to researchers and government agencies for developing scenarios or projections of future human demand on environmental resources.

Information Transmission Systems

Research on the characteristics and dynamics of environmental information transmission will allow government agencies and others to develop more effective ways to inform and, given appropriate policy decisions, to influence individuals' environmental choices. Government agencies and firms often rely on consumer information in the form of public awareness campaigns, public service announcements, informational flyers, labeling and rating systems, and so forth, to inform and influence environmentally significant behavior. It is sometimes assumed that lack of information is the main barrier to action and that informing or educating consumers will automatically lead to desired choices and behaviors. However, environmental programs that rely on disseminating information about behaviors and how to perform them have been notoriously ineffective on average for promoting the desired behaviors (Ester and Winett, 1982; Geller, Winett, and Everett, 1982; Hirst, Berry, and Soderstrom, 1982; National Research Council, 1984; Gardner and Stern, 2002). The results have been disappointing in part because information is often not the only significant barrier to action (Lutzenhiser, 2002; Gardner and Stern, 2002). But information has not achieved its potential in part because it has not been disseminated effectively.

More than two decades of research have identified a number of aspects of informational messages that can often be made better. For example, information is more effective when it is understandable to audiences, presented in a way that attracts attention and stays in memory, and delivered at a time and place close to the relevant choices (National Research Council, 1984; McKenzie-Mohr and Smith, 2000; Valente and Schuster, 2002). Important questions for message design include: How can information be made decision-relevant? and How should messages be designed to suit audiences' attention patterns, cultural understandings, and cognitive capacities?

Information programs can benefit by sponsoring applied research that addresses such questions in the context of their particular target decisions and audiences, and organizations that operate such programs should sponsor this kind of research. However, a general federally supported research program on message design is not needed.

Greater value can be gained from new research on the design and functioning of information communication and transmission systems. This research would take into account available knowledge about how information use is shaped by characteristics of information sources and their interactions with information producers and potential users. The significant characteristics include trust in information sources (National Research Council, 1984); the presence and behavior of intermediaries between information and its audiences (National Research Council, 1984); the use multiple information sources, including direct personal interactions (e.g., Werner and Adams, 2001; Mileti and Peek, 2002); the multiplicity of audiences that require different kinds of information and that trust different sources (Gardner and Stern, 2002); and the presence in most information environments of multiple, sometimes conflicting, sources and messages that purport to offer useful information (National Research Council, 1989, 1996).

The research would seek to develop and evaluate interventions in information transmission systems that take the above insights into account in order to get accurate and trustworthy information to a variety of audiences from sources they trust. Studies might examine environmental monitoring systems, production standards programs, certification schemes, and so forth to analyze their information transmission systems. They might consider how such systems and programs establish the trustworthiness of information about the environmental characteristics of goods and services (agricultural production practices used, resource depletion involved, effluents and their environmental impacts, habitat transformation, etc.), validate the information, and convey it to audience groups. Researchers might examine the roles of intermediary groups that have tried to provide information in forms that suit audience needs, to make it more readily available, or to increase its credibility to target audiences. An example is the creation of a web site by Environmental Defense to make information from the federal government's Toxics Release Inventory more readily interpretable by nonexperts (Herb et al., 2002). It is important to better understand the strengths and weaknesses of those approaches, how they are being used, and how their design and delivery can be improved.

It is also important to investigate more systematically other aspects of information transmission systems. One of these is the role of commercial advertising claims in environmentally significant consumer purchase decisions, both directly and in relation to public-sector information transmission. Another is the "reverse flow" of information—from consumers to firms, governments, and intermediaries regarding the kinds of information they need for their decisions, as well as about their desires in terms of products, environmental performance, and public policy. Yet another is the potential to design informational efforts for use when events make environmentally significant attributes of consumer behavior especially salient, as, for example, during the large spike in gasoline process in the spring of 2004.

The recommended research could fruitfully address general questions such as these:

- How effective are existing environmental information delivery systems?
- What are their limitations as primary sources of information?
- What information is most effectively transmitted by what sorts of intermediaries?
- How can information delivery systems (messages, intermediary networks, systems that produce indicators, etc.) be designed to produce and transmit information better suited to their consumers?
- How do public-sector information and private-sector marketing combine to shape individuals' awareness and understanding of the environmental implications of their choices?

Such questions might well be explored in the context of particular complex information transmission systems, such as those that lead to the labeling of produce as organic or as grown in accordance with ecological principles, or those that create green investment funds. The goal would be improved insight into the most appropriate roles of public, private, and nonprofit sector organizations in providing trustworthy information to a diversity of individuals, for example, as part of systems of environmental monitoring, certification, and the like.

Integrating Information with Other Policy Instruments

Research on the combined performance of information and other policy tools can significantly increase the effectiveness of all of these tools. It has long been recognized that information has limited value when there are weak incentives for its recipients to use it; also long recognized but not as universally acknowledged is the insight that incentives can fall far short of their practical potential if they are not accompanied by appropriate information about how best to take advantage of them. Several studies in the areas of consumer policy for energy conservation support this general insight (e.g., Stern et al., 1986; Brown, 2001; Brown, Levine, and Short, 2001). Environmental policy could benefit from improved knowledge of how information can be effectively combined with other policy instruments.

A number of governments and firms actively promote pro-environmental policies, products, and services—ranging from recycling programs and storm water management initiatives to energy efficiency technologies and alternative transportation modes. Policy strategies that have been developed to accomplish these goals include incentives, rebates, tax benefits, cost sharing, prohibitions, regulations, codes, standards, and the provision of infrastructure. The information component to such policies and programs involves making people aware of program opportunities and requirements, providing general guidance and detailed technical assistance on adoption or compliance and showing how particular patterns of voluntary choice will contribute to desired personal, social, and environmental outcomes. Information has been instrumental in changing individual behavior when coupled with public and market incentives, particularly in times of crisis (for an example in an energy shortage, see Lutzenhiser, Kunkle, Woods, and Lutzenhiser, 2003) and in a few intensive policy initiatives, such as the Hood River Project in Oregon, which attempted to bring all the homes in a community up to high standards of energy efficiency without regulation (Hirst, 1987). And as already mentioned, nonincentive factors associated with information transmission can produce a tenfold difference in the effectiveness of some incentive programs (Stern et al., 1986). Generic knowledge about ways to combine information with other policy instruments comes from studies in environmental and nonenvironmental policy arenas, the latter including public health (Valente and Schuster, 2002) and disaster preparedness (Mileti and Peek, 2002).

Useful applied research can be done in a variety of environmental policy contexts in which information and other influences are often combined. These include programs to encourage recycling (U.S. Environmental Protection Agency, 1994), energy-efficiency investments (e.g., Stern et al., 1986; Brown, 2001; Brown et al., 2001), changing behaviors involving household use of toxic chemicals (e.g., Werner and Adams, 2001), and transportation alternatives to the private motor vehicle (e.g., Brown, Werner, and Kim, 2003; Katzev, 2003). Research questions focus on how information can be supplied in complex policy contexts that allow persons to assess the social significance of their individual actions, effectively consider their policy and market options, alter their choices in concert with the actions of other individuals and organizational actors, and participate effectively in both markets and public-sector decision-making processes.

This applied research often involves the development and evaluation of policy innovations; policies are typically designed by focusing on a single tool, such as regulation or financial incentives, whereas these policies involve the coordination of tools. This research therefore presents two unusual challenges. One is methodological, involving the evaluation of an intervention that depends on the combined effect of several policy elements. The other is a challenge of dissemination that arises because policy innovations cannot necessarily be transferred in their entirety from one setting to another but may need to be adapted. Researchers and research users must find ways to make adaptations that will work in new situations. Such acts of diffusion of policy innovation can be an important object of research in their own right.

Fundamental Understanding of Consumer Choice and Constraint

A basic understanding of how information, incentives, and various kinds of constraints and opportunities, in combination with individuals' values, beliefs, and social contexts, shape consumer choice in complex real-world contexts would provide an essential knowledge base for understanding, anticipating, and developing policies for affecting environmentally significant consumer behavior. As already noted, environmental effects of individual and household consumption decisions are limited because consumer choice itself is seriously bounded, situated, and constrained by properties of the physical infrastructure, the range of options available in markets, legal and policy strictures, economic and information costs of behavior change, disposable income, household dynamics, and other factors. It is also influenced by commercial advertising, social comparison processes, and a variety of other social and economic forces.

Which of these contextual factors is most important, and the importance of contextual factors in general relative to such personal factors as values, attitudes, beliefs, skills, and information, depends on the type of behavior, its context, and characteristics of the sample of people studied (e.g., Black, Stern, and Elworth, 1985; Guagnano, Stern, and Dietz, 1995; Tanner, 1999; Gatersleben, Steg, and Vlek, 2002; Brown et al., 2003; Tanner, Kaiser, and Kast, 2004). The effects of information or other policy instruments on individual behavior are also dependent on a variety of factors. Thus, for example, to use information effectively, either alone or in combination with other policy instruments, it is important to understand what the possibilities for change are for target populations and what useful roles information of particular kinds can play. To know this, a more thorough knowledge is needed of the nature of real-world choice in messy cultural contexts.

A small but growing number of empirical studies are beginning to illuminate the determinants of several kinds of environmentally significant consumer choices. Some of them have used aggregate scales of pro-environmental behavior (e.g., Kaiser, 1998; Nordlund and Garvill, 2002; Cottrell, 2003). Others focus on particular types of consumer behaviors, including household energy use (e.g., Black et al., 1985; Poortinga, Steg, and Vlek, 2004), travel behavior (e.g., Bamberg and Schmidt, 2003; Katzev, 2003; Brown et al., 2003), food purchases (e.g., Tanner, Kaiser, and Kast, 2004), and recycling (e.g., Guagnano et al., 1995; Li, 2003; Oom do Valle, Reis, Manazes, and Rebelo, 2004). Some researchers have compared the determinants of different classes of pro-environmental behavior or have examined similar behaviors in different settings (e.g., different countries, where the contexts for such behaviors as recycling and mass transit use are dramatically different) (see, e.g., Stern, Dietz, Abel, Guagnano, and Kalof, 1999; Aoyagi-Usui, Vinken, and Kuribayashi, 2003; Thøgerson, 2004).

The findings so far suggest a few general principles, such as that the relative importance of contextual factors vis-à-vis personal ones is positively correlated with the environmental impact of the behavior (Black et al., 1985) and that the effect of personal variables, including information, is likely to be greatest when contextual pressures for or against a behavior are weak (Guagnano et al., 1995). Some studies suggest that behaviors that are grouped according to certain dimensions of similarity may have similar determinants (e.g., Stern et al., 1999; Thøgerson, 2004). But such generalizations have yet to be adequately tested across behaviors and settings. Additional research that looks systematically across behaviors, settings, and populations could considerably improve understanding of where the possibilities lie for information or other interventions to influence environmentally significant behavior.

Research also highlights the importance of at least four major classes of influences, attitudinal factors, contextual forces, personal capabilities, and habits (Stern, 2000). Researchers have been exploring finer distinctions within most of these broad categories, with the greatest amount of research attention being paid to the attitudinal class of variables, which includes basic values, environmental attitudes, identification with nature, beliefs about the environmental consequences of behavior, and other factors. Increasingly, researchers are developing coherent theoretical accounts of environmentally significant individual behavior that consider the roles and interactions of various kinds of influences and the possibility that different behaviors may respond to different collections of influences (e.g., Ölander and Thøgerson, 1995; Dahlstrand and Biel, 1997; Stern, 2000; Vlek, 2000; Gardner and Stern, 2002; Bamberg and Schmidt, 2003; Schultz, Shriver, Tabanico, and Khazian, 2004; Thøgerson, 2004).

An important research area is the apparent paradox already mentioned of strong environmental concern expressed in surveys of U.S. public samples combined with trends toward increasingly high environmental impact consumer choices, such as the purchase of larger homes and motor vehicles. Do these findings mean that some people's expressed environmental concern is only skin deep? That people do not understand the environmental implications of their purchases? That advertising has drawn consumers' attention away from the environmental attributes of products? That nonenvironmental product attributes are more important to consumers than environmental ones in their major environmentally significant purchases? That public policy and private business routinely deliver less environmental benefit than people want, in some sort of market failure and democracy failure? That consumer behavior and citizenship behavior are driven by fundamentally different psychological processes? Unraveling the paradox would provide valuable information to public- and private-sector decision makers concerned with the environmental impact of consumer choice.

It will be important for research effort on such questions as the above to build bridges across concepts and insights currently segregated by social science disciplinary boundaries. The dynamics of choice have been extensively studied by psychologists, anthropologists, sociologists, organizational analysts, and, more recently, behavioral economists (Smelser and Swedberg, 1995; Wilk, 1996; Bagozzi, Gurhan-Canli, and Priester, 2002; Camerer, Loewenstein, and Rabin, 2003). Their respective literatures are extensive, yet researchers in each area are, for the most part, uninformed about salient work in the others. The disciplinary literatures are also rendered somewhat alien to one another by differences in interests, paradigms, data, methods, and analytic vocabularies. However, these impediments to synthesis around a unified model of real-world situated choice could be overcome if significant funding were targeted to interdisciplinary inquiries in which just such a synthesis was the required outcome.

We stress the fact that the needed work is inherently interdisciplinary, which is also to say that work from any single perspective, even when aware of findings from cognate areas, will tend to fall victim to biases or blind spots that may result in misleading characterizations. The interdisciplinary work needed would best be done through the exploration of specific choices in particular settings, with an aim of developing policy-relevant knowledge on a number of questions: Which behaviors are most changeable? Which are most changeable directly by individuals? Which are most constrained by the choices of organizations or by societal infrastructure? How does this vary across social groups and technical circumstances? The applicability of research in specific settings to other settings can begin to be tested as bodies of knowledge develop in different settings, allowing a comparative analytic approach.

It will also be important to test generalizations developed in research in North America and Western Europe on consumers in other countries, particularly in developing countries where the most environmentally significant behaviors and the most significant incentives and constraints related to them are quite different from their counterparts in the high-income countries. Very little is known about environmental consumption in those countries at the individual and household levels. Understanding the forces affecting environmentally significant behavior in those contexts will be critical for projecting the environmental implications of consumer behavior in these countries overall and for designing policies for promoting green choices in those contexts.

Also little researched is the role of information in individuals' behaviors as citizens or as investors. Although it seems likely that the factors that shape individual environmentally significant choices differ when someone acts in different roles (e.g., consumer, citizen, or investor), little is known about this. Neither has there been much research on the relationships between individuals' consumer purchase or "demand" choices (e.g., paper versus plastic) and their actions and views regarding public policy. For example, the energy crisis in California in 2000-2001 showed that persons were willing to do their part by conserving electricity, but also that they expected significant visible conservation from businesses and aggressive government response to perceived gouging by energy suppliers (Lutzenhiser et al., 2003).

RATIONALE FOR THE SCIENCE PRIORITY

Likelihood of Scientific Advances

In all four areas identified under this science priority, past research has led to concepts and insights that open the way for further research. The opportunities lie both in fundamental research on consumer choice and in applied research focused on specific problems of measurement or of policy design and evaluation. The recommended research is likely to lead to understanding of broader applicability to issues of policy analysis and to fundamental understanding of individual behavior under complex real-world conditions. The availability of validated measures of environmental consumption would open a major avenue of research on the determinants of change in consumption.

It is worth noting that the geographic center of research on environmentally significant individual behavior, which was in the United States in the 1970s and 1980s when government interest in energy conservation provided research funding, now seems to lie in Western Europe. Even U.S.-based journals that publish in this area are increasingly dominated by non-U.S. researchers. The most likely explanation of this shift is the scarcity of research funds from U.S. sources for this research. An infusion of research funds has a high likelihood of turning the continuing interest in environmental topics among senior and young social scientists into significant contributions to policy-relevant knowledge for U.S. decision makers.

Potential Value

Research on information transmission systems and on developing and assessing mixed policy instruments would have obvious value for policy makers who want to influence individual behavior with information or other policy instruments. Improved basic understanding and measurement of environmental consumption would have significant value for environmental forecasting and modeling by improving on such common proxy measures as income, which do not account for behavioral differences among people with similar resources. It could also aid in anticipating the likely environmental effects of increased consumer spending in developing countries. Fundamental research also has longer term practical value by building a stronger base for applied research and policy-related understanding.

Likelihood of Use

In each of the four research areas identified, a variety of potential users of the results of the recommended research are likely to put the findings to use in policy and program development. It is important to note that state and local governments, which are often more dependent on voluntary citizen or consumer action by itself or in conjunction with regulation, may find this work particularly useful when applied to local circumstances and resource constraints.

Trustworthy indicators of consumption would be used by individuals as consumers and investors; government agencies at various levels that have environmental education mandates or that need to forecast consumer demand on environmental resources; producers and vendors of green products and services; and nongovernmental (e.g., environmental and consumer) organizations, green investment advisers, and firms concerned with environmental accounts. The extent of use by consumers will depend on how the information in indicators is delivered and on concurrent policy and market incentives and constraints on choice. Credibility is also likely to be an issue. Cooperation among researchers, environmental agencies, intermediaries, and likely users is desirable for making indicators useful and credible.

Improved understanding of information transmission systems would be useful for government agencies that require or provide environmental information, particularly to achieve environmental goals that may be difficult to achieve by regulation or economic policy instruments; firms that supply environmental information and have reputation concerns; and trade associations, investment advisers, and nongovernmental organizations that might act as intermediaries or monitors in complex information transmission systems. Organizations concerned with designing effective environmental certification systems or credible green investing instruments would be quite likely to apply findings from such research. Government agencies might also apply these findings when designing information production and delivery systems and when evaluating the likely success of privately developed information programs (such as those presented as alternatives to regulation). Use of this research will depend in part on the degree of pressing need for credible information transmission systems. Strong pressure for effectiveness in information-based policy will also increase the likelihood of use. If credibility scandals arise with existing information systems, such as those based on information voluntarily provided by polluters, many people will wish that research on information transmission systems had been conducted sooner.

Improved understanding of how information can be integrated with other policy instruments would allow governments to apply this knowledge to design more effective policies for behavior change that combine public information, market incentives, infrastructure investments, and other interventions. In California, for example, serious consideration is now being given to promoting large-scale energy usage changes through a combination of public interest advertising, consumer incentives, research and development, and codes and regulations. A number of cities are involved in similarly ambitious efforts—e.g., addressing a variety of air quality, water quality, traffic congestion, solid waste, wastewater, and infrastructural capacity issues through combinations of prohibitions, inducements, public information, and voluntary civic action. They would use the results of this research if some of it is conducted in the context of their programs and policies and if research results are actively disseminated through relevant intermediaries, such as associations of state and local government officials.

Improved fundamental understanding of the dynamics of choice and constraint would be useful to environmental forecasting enterprises and the designers of environmental policy instruments: firms and governments concerned with strategic targeting of messages, incentives, and products. It would provide policy makers with background information regarding the relative potential for informational and other policy instruments both alone and in combination; the limits of informational interventions and the places where barriers would need to be removed if information is to be effective; the kinds of information that should be targeted to different types of audiences (e.g., individuals acting for themselves, intermediaries, people acting on behalf of the organizations). In combination with the development of environmental consumption indicators, this research would help in setting priorities so that attention is given first to behaviors that are both changeable and worth changing. Knowledge of the fundamental dynamics of choice would also have applications in nonenvironmental areas concerned with the effects of choice on economies, polities, health and welfare systems, etc.

The use of the results of the recommended research is likely to depend significantly on the quality of communication between researchers and the potential users of their findings. Particularly because much of the recommended research would be locally based, communications networks will be important for promoting the diffusion of insights from such research to other places where they might be applied. Government agencies concerned with the widespread application of the recommended research should therefore also consider providing support and encouragement to efforts to establish ongoing communication between the producers and potential users of research pursuing this science priority. Such efforts might be organized by existing organizations, such as consumer groups, trade associations, or associations of local governments, which might create networks of researchers and practitioners, invite researchers to meetings of possible research users, or develop other innovative communication mechanisms.

6

Decision-Relevant Science for Evidence-Based Environmental Policy

To strengthen the scientific infrastructure for evidence-based environmental policy, the federal government should pursue a research strategy that emphasizes decision relevance. Such a strategy should include four substantive research elements: (1) developing decision-relevant indicators for environmental quality, including pressures on the environment, environmental states, and human responses and consequences; (2) making concerted efforts to evaluate environmental policies; (3) developing better methods for identifying the trends that will determine environmental quality in the future; and (4) improving methods for determining the distributional impacts of environmental policies and programs. These research elements require contributions from both the social sciences and the natural sciences, as well as communication across scientific communities.

Because scientific information is critically important for environmental decision making, major research efforts in environmental science are often justified by their value to society. These efforts typically produce high-quality science, but they have repeatedly fallen short in addressing the questions most important to societal decision makers (Oversight Review Board of the National Acid Precipitation Action Program, 1991; National Research Council, 1996, 1999b, 1999d, 2001a, 2004c): they have failed to produce the right science for decision making. By pursuing this science priority, federal agencies would greatly improve the infrastructure of scientific information and methods relevant to addressing questions of importance to the potential beneficiaries of environmental science. The recommended research would allow environmental decision makers better to understand the nature, severity, and causes of environmental degradation, to learn from the substantial experience of designing and implementing environmental policy in the United States, and to anticipate environmental trends and future policy needs. By focusing scientific efforts increasingly on decision relevance, such a program of measurement, evaluation, and analysis would increase the influence of empirical evidence and empirically supported theory in environmental decisions relative to the influences of politics and ideology. It would integrate the social sciences and the natural sciences of the environment and build a knowledge base that would better inform practical decision making while also informing scientific research.

Processes for determining which research is most decision-relevant should be participatory. Choices about how to construct indicators, evaluate policies, and so forth should be made with the participation of the full range of likely users of measures, evaluations, and analyses. This approach has previously been recommended for shaping scientific research agendas in a number of disparate areas of environmental science (e.g., National Research Council, 1996, 1999d, 2003, 2004c), and we state it here as a general principle. Broad involvement is essential to enhance the decision relevance and credibility of measures, evaluations, and analyses. Choices about measurement are not purely technical, so they are not appropriately left to analysts alone. Measurement focuses attention on what has been measured and away from what is unmeasured, thus embodying values about what is most worthy of attention. The affected parties in environmental decisions often disagree about what is most worth measuring, which outcomes of policies are most important, and the like, and the range of measures needed to make an analysis credible may not be obvious to scientists or government officials unless the various potential users of indicators are involved. Choices about what evidence to collect for policy are probably most appropriately made through broad-based analytic-deliberative processes, such as described in Chapter 2.

BUILDING THE INFRASTRUCTURE

Government should implement the strategy of seeking decision relevance in each area of environmental policy. This will require four kinds of research activities: (1) developing decision-relevant indicators, (2) evaluating past policies and programs, (3) improving the scientific capability to anticipate future environmental conditions and problems, and (4) measuring and monitoring the distribution of environmental impacts in the population in relation to issues of environmental inequities and their abatement. Within each of these activities, decisions about research priorities should be informed by dialogue between the potential producers and the potential users of the research.

Evidence-based environmental policy depends on having measures and analyses of all the important decision-relevant aspects of the systems that policy affects. Meeting this need may require departures from usual routines of environmental measurement and analysis in at least three ways. First, the focus of measurement must encompass pressures on the environment, environmental states and conditions, and the human consequences of and responses to those conditions. Measuring conditions of the biophysical environment in isolation from its human interactions is insufficient because to anticipate the need for policy action or to anticipate or assess the results of any policy choice, it is necessary to consider the conditions of both the human and nonhuman parts of the system. Thus, it is important to measure and analyze the environmental implications of human actions that are taken for nonenvironmental reasons (e.g., trade, technological and economic development, national security, and so forth), which can exert major pressures on environmental systems or shape human responses to environmental conditions.

Second, evidence-based policy depends on an appropriately linking the social sciences and the natural sciences of the environment to provide the needed measures and analyses. The social and behavioral sciences are critical for developing measures of human influences, consequences, and responses, as well as for developing and refining processes for selecting useful measurements.

Third, evidence-based policy requires the participation of both the likely producers and the likely users of the evidence in deciding which measures and analyses are needed. This is so because for information to be decision-relevant, it must serve the needs of a variety of decision participants outside the scientific community, as well as those of scientists. In some environmental decision contexts, the potential information users differ considerably in the issues that concern them and in the kinds of information they want. Because there are often too many issues of concern to measure and analyze them all, difficult choices must be made about what, when, and how to measure. If these choices are not informed by the perspectives of a sufficient variety of information users, then environmental analysis is likely to fail to provide an adequate evidence base for decisions. Thus, the purpose of engaging information users is to promote the accuracy, rigor, decision relevance, transparency, and credibility of environmental information and analysis. The roles of the various participants should be defined to promote the achievement of these goals.

DEVELOPING DECISION-RELEVANT INDICATORS

Social science and natural science research should be integrated in a comprehensive approach to developing indicators that are relevant and usable for environmental policy. These indicators should cover not only

states of the biophysical environment but also human influences on nature and the impact of the physical world on humans. People involved in decisions that affect the environment need information to help them understand the ways in which possible decisions may affect pressures on the environment and for anticipating and assessing the effects of decisions on things people value. To produce such information requires integrating natural science and social science approaches, involving both the potential producers and users of information and engaging the relevant government agencies.

Measurement is the heart and soul of scientific inquiry and is also essential in defining, understanding, and managing human and environmental affairs. It is necessary for forecasting and planning for environmental situations and for assessing the results of decisions taken. Therefore, an integrated approach to measurement recognizes that environmental measurement has important users outside the scientific community and that systems of environmental measurement must serve the needs of those users as well as those of scientists. Because these various constituencies differ in the environmental situations that concern them and in the kinds of information they want, and because there are too many aspects of human-environment interaction and too many things potentially affected by policy to measure them all, difficult choices must be made about what, when, and how to measure. It is important to involve both the producers and the various kinds of users of environmental information in making these choices.

Indicators are quantitative measures, collected and reported on a regular basis, that convey useful information. Environmental indicators are based on data, such as from environmental monitoring systems, but not all data convey useful information. Thus, the minute-by-minute readings from a continuous air pollution monitor are data that can be aggregated into indicators, such as a daily average pollution level. Indices also convey useful information but are often more widely aggregated to represent broad conditions at the moment and through time. For example, the Dow-Jones Industrial Average on the New York Stock Exchange is an economic index having enormous general decision-making influence, although no one claims that it represents the precise details of any firm or sector of the economy. It is a useful composite or proxy when there is too much information to understand. An air quality index such as the U.S. Environmental Protection Agency (EPA) has developed may combine information on various pollutants measured at different locations. There is some overlap among terms; the same numbers may be data, indicators, and indices.

Indicators developed for decision relevance would enable major advances in fundamental understanding of the dynamics of human-environment interaction by vastly increasing the possibility of analyzing these relationships quantitatively. They would also greatly increase the decision relevance of environmental analyses that use them by providing credible measures of variables of critical concern to both decision makers and scientists.

The Research Need

Indicators are necessary for rationally formulating, implementing, and evaluating environmental policy. In most situations, indicators are the only environmental reality that decision makers see because the environment itself is too vast and intricate to be perceived directly. Indicators are the institutional sense perceptions that tell us about the environment. The essential function of indicators is reflected in the fact that almost all environmental agencies conduct some kind of monitoring on which to base indicators. The federal government is estimated to spend $500-600 million annually on environmental monitoring. Unless the monitoring data are incorporated into indicators, they are not likely to be useful. There is no environmental equivalent to the Dow Jones; however, it is possible to go well beyond the current measures of biophysical phenomena by adding information about critical human dimensions seldom taken into account in conventional measures.

Environmental indicators have a variety of uses and users. They may be used for research, policy formulation, enforcement, management, evaluation, and public information. Users can include scientists, policy researchers, public officials at all governmental levels, representatives of affected parties, the mass media, and the public. Until now, approaches to developing indicators have not followed a comprehensive, integrated approach organized by the need to inform decisions. Rather, they have been fragmented by academic discipline, government agency, geographical and temporal scale, and in other ways. They have been driven by the perceived needs of specific groups, such as scientists and government officials responding to legislative mandates. Partly as a consequence, no set of environmental indicators in the United States commands the respect and attention of the public or policy makers.

The users of indicators typically want to know not only about environmental conditions, but also about their human connections. The Organisation for Economic Co-operation and Development (OECD) accordingly distinguishes three types of indicators: *pressures* (e.g., population, technology, consumption, pollutant emissions); *states* (e.g., the condition of ecological areas and biota); and *responses* (public and private actions taken to reduce pressures, protect states, and adapt to environmental changes). An indicator system might also include measures of the human consequences of environmental events that take responses into account. EPA's 2003 Draft Report on the Environment (p. viii) proposes a six-level hierarchy that roughly parallels the OECD framework, although it differs by making it a hierarchy rather than a process with feedback loops. Also, the EPA effort was based on a narrower definition of environment that neglected such matters as resources and energy.

Environmental indicators in the United States have usually been framed narrowly. Initially, the effort to develop indicators was dominated by statisticians. The result was statistically sophisticated indicators that paid little attention to usefulness, either in terms of communicability or the needs of decision makers. Indices that combined diverse types of data were generally frowned on because, in the view of many statisticians, too much detailed information was lost when the data were combined into indices.

Ecologists have dominated the two most recent indicators efforts, the Committee to Evaluate Indicators for Monitoring Aquatic and Terrestrial Environments (National Research Council, 2000) and the Heinz Center group that received a mandate from the White House to develop an "environmental report card" (Heinz Center, 1999). These efforts have proposed indicators of environmental states but have omitted indicators of pressures and responses, even though, as the National Research Council report noted, indicators of pressures are "no less important" than indicators of states (p. 2).

Such efforts have not dealt adequately with measures of pressures or responses, although there are ample data that could be used to produce indicators of them. For example, data on materials flows (National Research Council, 2004a) can contribute to indicators of environmental consumption (see Chapter 5), a major pressure variable. On the response side, federal agencies have developed ample data because of the requirement under the Government Performance and Results Act of 1993 that all federal agencies measure how their missions are being accomplished. Most of these data lack any connection with measures of pressure or state, and so they are much less useful for evaluation purposes than they might be. Health data may provide another source of environmental indicators on the response side (see U.S. Environmental Protection Agency, 2003:4-1 to 4-24). The role that social scientists have played in risk assessment shows that they may make useful contributions to delineating the environment-health link (e.g., Krimsky and Golding, 1992). Efforts to modify economic indicators to take environmental factors into account have the potential to yield useful indicators related both to pressures and responses (e.g., National Research Council, 1999c). Economists have contributed much to this effort and can contribute more.

An integrated approach to decision-relevant environmental indicators will benefit greatly from the involvement of social scientists. They can bring expertise regarding the measurement of pressures, responses, and the human consequences of environmental events, as suggested above, as well as techniques and experience for modeling the linkages in complex systems. They thus can contribute significantly to an integrated system that includes pressure, state, and response measures. Such a system may require new mixes of disciplines, likely to include natural scientists, social scientists, and engineers.

Integrating social scientists will take special efforts. Until now, they have played almost no role in the development of environmental indicators, with a few exceptions, such as efforts to develop a green gross domestic product (National Research Council, 1999c), quality-of-life indicators (National Research Council, 2002c), and indicators of environmental sustainability. Despite considerable interest in results-based or outcomes-based environmental governance, a new book on *Environmental Governance Reconsidered* (Durant, Fiorino, and O'Leary, 2004), with essays by the leading social scientists in the field, does not contain any discussion of indicators to measure results or outcomes.

To ensure the decision relevance and comprehensibility of indicators, government agencies involved in developing them should create them in collaboration with the producers and potential users of the information, including a variety of nonscientists. The environmental indicator movement has run into trouble by asking scientists and almost no one else what is important to measure. This narrow approach is both common and problematic with science-based efforts to inform decisions, such as risk assessment (National Research Council, 1996), climate forecasting (National Research Council, 1999a), radioactive waste management (National Research Council, 1995), and valuation of biodiversity (National Research Council, 1999b). Indicators need to be built on an understanding of what matters most to the interested and affected parties to decisions (National Research Council, 1996), not just what matters most to scientists. Indicator development needs also to recognize that information that might be meaningful to one individual in given circumstances may be irrelevant or incomprehensible to another person elsewhere. For example, indicators that are meaningful to an ecologist studying an aquatic environment may be incomprehensible or of questionable relevance to a fishing boat operator.

Decision-relevant indicators must be developed with due attention to the variety of decision participants and the range of values that matter in specific decision-making settings or contexts. Indicators need to be understandable, quantifiable to the greatest extent possible, applicable to the realistic setting or circumstances at hand, and relevant to users' concerns. They must portray information about the past and present of valuable ecosystems and include relevant and significant human dimensions all at the same time. Good indicators give a picture or map that reveals fundamental issues and suggests possibilities. They allow one to extend past trends (what forecasters and planners call the "null projection") to generate a plausible scenario for the future. They make problems stand out in strong relief and suggest where analytic and managerial attention might best be focused.

Indicators often reflect values in the sense that people make judgments about which direction they would like the indicator to go (e.g., less air pollution is better, fewer species is worse). This value component makes it important to show that the link between the indicator and the relevant value is really valid. For human health indicators, the validity problem involves finding the right weights to give to diverse health threats so they can be combined into a single scale. For ecological indicators, a problem is that concepts of ecosystem health are very subjective (National Research Council, 2000:24). Social scientists can contribute much to the clarification and amelioration of these problems, for example, by studying people's understandings of proposed indicators. Decision scientists can help by encouraging other scientists to distinguish the variables that may be elements of human or ecosystem health in ways that promote empirical analysis.

Decision participants' values also reflect their geographic, socioeconomic, cultural, occupational, and ethnic positions. Because of these different positions, some individuals face greater risks than others from particular sources of pollution or types and locations of resource degradation, and their exposures to risks and opportunities affect the kinds of information they need most for their decisions and the kinds of indicators they find useful and credible. Spatial decision support tools are being developed that can help illuminate the spatial and temporal aspects of particular environmental pressures or conditions, such as climate change, flooding, habitat degradation, and the transport of pollutants, but far less is known about how to illuminate the socioeconomic or cultural aspects of the uneven distribution of environmental risks and benefits across human populations. The development of decision-relevant indicators should take into account diversity among information users with regard to which information they need most.

For such reasons, choices about what to measure and how to aggregate data into indicators raise important and researchable questions about strategy and procedure. The answers will be contingent on the nature of the problem, decisions to be informed, the scale of those decisions, and other factors. Because the users of indicators normally have various perspectives on these matters, the answers may also be contingent on who asks the questions.

These choices should therefore be informed by the needs of potential users of indicators. The costs of environmental measurements need to be balanced against the value of information they provide. Scientists have a hearty appetite for data, but from a decision-making perspective, it is important to strive to make indices few, inexpensive, comprehensible, and decision-relevant. Using and modifying the data collection and analysis efforts of others is one tactic to attain this end. Assessing the value and utility that decision makers and the public place on various indicators and discarding the least effective is another.

Users' concerns have led to interest in issues of environmental justice, especially in the context of evaluation of environmental policies. Although there is considerable debate over what constitutes environmental injustice (Bowen, 2002), such debate can be informed by indicators that represent the consequences of environmental decisions at sufficiently fine-grained scales to identify how environmental risks are distributed among different populations. We return to this issue later in the chapter.

Good indicators must be not only technically valid and decision relevant, but also comprehensible to the potential users, who include a variety of nonscientists. Broad indices that combine several different types of data have especially strong potential for making information useful to the public and the mass media. Little work has been done on indices, however. There is a need to balance the concerns of technical specialists familiar with the environmental data, who often feel that indices distort the individual data sets, and the need to make information broadly meaningful. The EPA's effort to develop air pollution indices (e.g., U.S. Environmental Protection Agency, 2003:1-4) is important but exceptional. There is, for example, no equivalent effort to develop water pollution indices.

Social scientists have a good deal of expertise that can be brought to the problem of index construction, including expertise in communicating information and in combining data from diverse sources. They also can help in analyzing the strengths and weaknesses of indices, such as the sustainability indices that have been proposed.

The objectivity or disinterestedness of the parties collecting and presenting indicators cannot be overemphasized. Transparency of collection, measurement, and presentation is crucial, particularly when trying to earn and maintain public trust and confidence. The essential questions here are: Who should provide the information? How should objectivity and transparency of the work be ensured? How should the information be made accessible to a full range of participants?

Good indicators require close collaboration among existing organizations and may require the creation of new ones. Many data problems arise from a lack of coordination among different institutions. Federal agencies often do not share data with each other, data collected by states may not be in a form useful to the federal government and vice-versa, and data sharing and review between the public and private sectors may be inadequate.

Data improvement also may depend on the creation of new institutions. There have been repeated attempts to create a federal Bureau of Environmental Statistics but none have succeeded. Environmental policy is thus deprived of a source of impartial data and indicators. Almost all other major policy areas have an institution that supplies this need, such as the Bureau of Labor Statistics, the Center for Health Statistics, and the Energy Information Administration. Recommendations to create a federal Bureau of Environmental Statistics deserve serious attention because of the need for collaboration. There is also a need for stronger institutions to provide international environmental indicators.

Social scientists know a good deal about improving coordination and about creating and strengthening governmental institutions. They could thus help to strengthen the environmental data base and lay the groundwork for institutions that will provide good environmental data on a continuing basis.

Special efforts may be required to enable rapid development of indicators under conditions of surprise or disaster. It is impossible to develop indicators for every environmental policy situation that may arise. It may be possible, however, to build the capability for developing indicators quickly when there is a new need. Consider the *Exxon Valdez* disaster. Even though the accident affected a highly valued and well-studied ecosystem, there were unanticipated needs for measurement and for rapid scientific judgments about what to measure. Existing environmental indicators certainly helped those making decisions about minimizing the impacts of the spill, but much more was needed—and fast. Had the *Valdez* simply sunk in the middle of the ocean, the need for environmental measurement would have been far less, but also far different.

For such unexpected environmental events—oil spills, natural disasters, nuclear accidents, terrorist attacks, and so forth—a measurement strategy of rapid assessment makes sense. In these situations there may be preexisting environmental data and information, but probably not enough and not in forms that decision makers need or can use. The scientific challenge is thus to bring existing general knowledge about processes and interactions to bear as a means of selecting and sorting through the entire array of possible things to measure in the specific circumstances.

The interesting scientific tasks here involve anticipating how the surprising events might develop and then imagining and deducing what information would be required to cope. In the *Valdez* disaster, ecological and economic consequences were among those needing measurement. In the wake of the destruction of the World Trade Center on September 11, 2001, the public health effects of exposure to tons of toxic dust were among the effects to be monitored. The medical concept of triage is relevant as a reminder to decide what is important and manageable. Of course, these decisions call attention to other questions: importance to whom and manageable at what costs?

Rapid assessment efforts can draw on the knowledge and experience of social scientists, particularly disaster researchers, in rapid mobilization of scientific efforts in disasters and in understanding the ways crisis scenarios typically proceed. Rapid assessment would require teams with expertise across the sciences to consider the full range of consequences to be monitored and indicators to be developed.

Efforts to develop indicators should include the following steps regardless of the environmental system or problem for which indicators are needed. In general such efforts should entail the following steps: (1) clear identification of the user audience and the uses to which the indicators will be put; (2) an inventory and evaluation of existing efforts and indicators; (3) development of new methods for indicator construction and new indicators, if necessary; (4) identification of the data needed for the indicators and an evaluation of the availability of the data; and (5) pilot testing of each indicator to analyze how well it meets the specified uses. Extensive trial-and-error testing of different indicators will usually be necessary to see which ones best met the needs they were intended to fill.

Rationale for the Activity

Environmental indicators that are designed as recommended, to be decision relevant, would clearly have great potential value for environmental decision making. They would also have a strong likelihood of being used, despite some resistance that might be anticipated. The federal government makes large investments in indicators now, and these may well increase because of growing emphasis on assessing the results of government policies (see the section below on evaluation). Thus, improved indicators probably would be welcomed by a wide variety of potential users, and particularly by environmental policy makers.

Less certain is the issue of whether the field is ripe for making significant advances. Part of the problem is that the kind of transdisciplinary research community that is needed for developing the needed kinds of indicators for decisions at the national, state, and local levels is not yet organized. There is no academic discipline or set of journals associated with developing the kinds of comprehensive environmental indicators needed for policy decisions in the United States. There are social scientists working on related kinds of indicators, for example indicators of sustainable development and of vulnerability to climate change, but relatively little of this work has been directed to developing indicators to address salient environmental policy questions in the United States. Thus, this field cannot be judged ripe based on preexisting research. But the existence of research communities working in closely related areas suggests that the area is ripe for development by attracting researchers from those areas.

An effort to develop the recommended kinds of indicators might be sufficient to attract those social scientists and create the kind of community of scholars that exists in other research areas. The scholars are there. The research questions exist. What is needed is the spark to bring them together, and a governmental effort to develop better environmental indicators, even for a single important environmental system, could provide such a spark. There is a wealth of environmental indicator material available for social science researchers to use, and the field is virgin territory. If policy makers convert their demand for answers into a demand for comprehensive indicators, this demand could mobilize a new set of researchers who have the necessary skills but lack only the recognition of the contribution their skills could make.

EVALUATING ENVIRONMENTAL POLICIES

Federal agencies should support a concerted research effort to evaluate the effectiveness of environmental policies established by public and private actors at the international, national, state, and local levels. This research would analyze the evidence of the effects of past policies by applying and adapting techniques of evaluation research that have been used to assess the effectiveness of social welfare policies to the domain of environmental protection. It would examine the outcomes of environmental policies and their alternatives in terms of such policy goals as effectiveness, efficiency, fairness, and public acceptability; strengthen methods and capacity for determining the results of environmental policies; and thereby help to answer the call for results-based management in government.

The Research Need

Due in large measure to the series of sweeping environmental statutes signed into law in the 1970s and 1980s, the quality of the nation's air, water, and land has improved dramatically over the past several decades. But even while these laws have led to substantial gains, many in government, business, and environmental advocacy organizations maintain that environmental policies could be more effective in achieving their goals (National Academy of Public Administration, 1995; Ruckelshaus and Hausker, 1998; Portney, 1990). Evaluation of environmental policies is a necessary step to improving policy effectiveness.

Using the tools of evaluation research, those in government and the private sector can begin to answer a range of pressing questions concerning the appropriateness and effectiveness of environmental policies. Policies may be evaluated before they are implemented (ex ante) to determine the "best" option among alternatives. "Best" may be determined on the basis of efficiency, as in the case of benefit-cost analysis, or by risk reduction, fairness, or other environmental, social, political, or economic criteria. Policies may be assessed after implementation (ex post) to determine whether any of a number of conditions has changed: public health conditions, environmental quality, environmental performance of regulated entities, or managers' perceptions about the costs, benefits, and effectiveness of the new rules. Evaluation can answer whether programs are accomplishing what they set out to do, where additional resources are likely to advance policy objectives, and where continued efforts are likely to prove fruitless. Evaluation allows policy makers to learn from experience: to identify the lessons of public and private experience developing and implementing environmental policy and to use these lessons to make policies more effective.

Despite the substantial potential value of rigorous evaluations of environmental policies, such ex post efforts are relatively rare (see Appendix D). Most evaluations of environmental policies are conducted ex ante and focus on whether the costs, measured in dollars, will outweigh the benefits (Knaap and Kim, 1998). Such assessments have the limitation of judging the appropriateness of a proposed policy on the basis of a single measure. Also, because they are essentially projections, they do not draw on evidence from actual experience.

Several factors make it difficult to evaluate environmental programs rigorously (Appendix D; Harrison, 2002). One is the need to judge the outcomes of a policy for the group or region that is subject to it against the outcomes for a suitable comparison group. When environmental policies are national in scope, finding such a comparison group is especially difficult and it is necessary to construct plausible counterfactual scenarios. Another obstacle is the lack of suitable outcome indicators against which effectiveness can be measured. The need for such indicators is discussed elsewhere in this chapter. It is worth noting that indicators developed for different purposes may not be right for policy evaluation (Gormley and Weimer, 1999). Also, when indicators are used for management, they have the potential to distort incentives and encourage managed entities to "teach to the test," as the phenomenon is called in educational evaluation.

A major technical obstacle to rigorous evaluation is the necessity and difficulty of controlling extraneous variables. To establish a causal connection between a policy action and change in the natural environment or human outcomes, it is necessary to extract from consideration major influences that are independent of the policy, such as fluctuations in the economic cycle, developments of new technology, changes in land use and development, meteorological conditions affecting pollutants and natural resources, and so forth (Powell, 1997). Although there are ways to address this problem technically, the extraneous variable problem, especially the problem of economic effects, has generally been ignored in environmental policy evaluation. Social scientists could make a major contribution by researching how to control for extraneous variables that muddy the interpretation of the effects of environmental policy decisions.

Evaluation of policies is also made difficult by the number of variables that shape how the targeted entities perceive and respond to policies. In firms, these variables include a facility's history of environmental management, size, customers, and numerous other factors (see Appendix D). With individuals, they include attention to messages, trust in the sources of information, and perceived difficulties of engaging in the behaviors the policies are promoting (National Research Council, 1984; Stern et al., 1986; Gardner and Stern, 2002). Research needs related to understanding the factors that shape environmental performance are discussed for firms in Chapter 4 and for individuals and households in Chapter 5.

Finally, there is the potential for evaluation research to be used to justify or delegitimate existing policies according to a government's political agenda. This potential underlines the importance of establishing effective quality control and review procedures for environmental policy evaluations.

Areas of Research

One useful area for evaluation research concerns the ongoing debate about the effectiveness of environmental regulations. For example, technology-based regulatory standards that specify how regulated entities must act and performance-based standards that specify the outcomes they must achieve have been credited with substantial improvements in environmental quality (Ashford et al., 1985; Houck, 1994; Shapiro and McGarity, 1991). Yet these approaches are also criticized for being rigid, fragmented, complex, costly, and failing to accommodate or motivate innovation (Ackerman and Stewart, 1985; Ayres and Braithwaite, 1992; Chertow and Esty, 1997; Fiorino, 1996; Gunningham and Grabowsky, 1998; Hahn and Stavins, 1991; Orts, 1995; Pildes and Sunstein, 1995; Stewart, 2001; Teuber, 1983). Furthermore, such regulatory approaches may be unsuitable for addressing environmental problems created bydisbursed, or nonpoint sources; environmental impacts from consumer, service, and agricultural sectors; or problems whose source is distant in terms of time or place. Other styles of regulation, such as by risk, exposure, licensing, and so forth, also have strengths and limitations. These issues raise various evaluation questions: What are the accomplishments and limitations of particular styles of environmental regulation at international, federal, state, and local levels? Do they address the most pressing environmental problems? What does each style of regulation do best, and what does it fail to do? What, in other words, are the comparative advantages of different regulatory approaches and the needs for innovation in environmental policy?

Evaluation can also focus on a variety of policy reforms that EPA and state agencies have initiated to address the judged deficiencies of past regulations. These reforms have been categorized into three broad types: informal rules, economic incentive systems, and reflexive law (Stewart, 2001). Informal rules allow regulators leeway in how they interpret environmental laws. For example, EPA has tailored regulations to fit the specific circumstances of a firm (e.g., Project XL) or an industry (e.g., Strategic Goals Initiative for the Metal Finishing Sector). Economic incentive systems impose a price on pollution and allow individual firms discretion in the level of pollution they generate (e.g., sulfur dioxide emissions trading programs). Reflexive law attempts to create conditions under which facility managers are exposed to new sources of information about their environmental conduct, so that they reflect self-critically about their performance (an example is the Toxics Release Inventory, which requires firms to disclose environmental performance information and better enables them to monitor and manage their own emissions). These new approaches have assumed substantial significance in parts of EPA and state and local environmental protection offices.

Should the existing legal structure change to accommodate alternatives to conventional environmental regulation? How do these innovations stack up against the status quo? To what degree do they address the deficiencies critics have noted? What are the conditions under which they seem most suitable? Answering such questions will require developing a clearer understanding of how environmental innovations have worked in practice. Some innovative policies have been subject to systematic evaluation, including market-based instruments (e.g., Tietenberg, 2002) and negotiated rule-making (Langbein, 2002). But many innovative approaches, even some that are mature programs, have so far not had the benefit of serious assessment. Such evaluation would be valuable for informed decisions about institutional and legal change.

Evaluation research should also focus on the impact of environmental policies on the behavior of firms (see Appendix D and Chapter 4). While EPA has been experimenting with innovative programs, many firms and trade associations have apparently moved from fighting external pressures for better environmental protection to incorporating environmental concerns into internal decisions and normal management practices. What is the link between environmental policies promulgated by agencies and these changes in the ways managers define their responsibilities? To what degree do the environmental policies of Europe and international organizations affect firm behavior in the United States? Fruitful areas for evaluation include the effects of the Coalition for Environmentally Responsible Economies, the American Chemistry Council's Responsible Care initiative, and the ISO 14001 system of third-party certification. The lessons of such efforts have not been learned or incorporated into policy.

Learning from evaluation can help public and private policy makers better determine the appropriate roles of various policy approaches. What shape should environmental policy take based on what we have learned from experiences with various styles of regulation and alternative policies? How are various approaches most appropriately combined in an overall policy strategy?

Establishing environmental policy evaluation as a research priority will encourage researchers to answer such questions. It will also strengthen methods for untangling the causal relationships between policy implementation, behavioral change, and environmental outcomes. These causal relationships are difficult to discern, whether the question at hand is the impact of a local land use policy on wetlands protection or the effectiveness of a federal initiative to conserve energy. Determining such causal relationships requires sophisticated understanding of how best to design research, utilize available data and overcome data gaps, and identify appropriate comparison groups. A concerted research effort on environmental policy evaluation will lead researchers to address these methodological challenges and advance the field. Such advancement should include expansion of evaluation criteria to include criteria of substantial importance to segments of the public, such as fairness and inclusiveness.

Rationale for the Activity

Evaluation research has a history extending back to assessment of the effectiveness of the New Deal programs of the 1930s (Rossi and Freeman, 1993). It became much more widespread after President Lyndon Johnson signed an executive order in 1965 establishing the Planning-Programming-Budgeting System (PPBS) as a requirement for federal program managers, leading to substantial increases in public spending on evaluation and the establishment of offices of program planning and evaluation in federal agencies (Haveman, 1970). Policy evaluation came to involve educators, sociologists, public health scholars, and psychologists, all bringing distinctive and useful measurement tools from their disciplines (Suchman, 1967). Later, researchers joined from economics and political science and then from various management disciplines. During the 1960s and 1970s, most policy evaluations focused on the effectiveness of human service programs to determine ex post whether efforts were achieving their intended results (Caro, 1971). The early generation of evaluation research, though often academically rigorous, was not necessarily intelligible, timely, or useful for decision makers. The simple question "Is this program working?" often was ignored or could not be answered (Szanton, 1981). The mismatch between analytic standards and client acceptance has long been a focus of thinking and writing about evaluation (Cronbach et al., 1980).

During this period environmental policies were subject to ex post review only rarely, due in part to the conceptual and methodological challenges already noted. Environmental policy evaluation was also impeded by the expense and difficulty of collecting outcome data, which often could be interpreted only through models that took into account meteorological conditions, human and environmental exposures, uptake mechanisms, and so forth.

Relevant environmental data are more readily available now than previously. For example, researchers can access the substantial information EPA and states collect from regulated plants to evaluate the effectiveness of regulations and alternative policies. This information is a valuable resource for program evaluation (Metzenbaum, 2003). Compliance and enforcement data about facilities, available in on-line databases such as EPA's Integrated Data for Enforcement Analysis (IDEA), Enforcement and Compliance History Online (ECHO), and the Sector Facility Indexing Project (SFIP), can be used to study trends in regulatory compliance and the relationships between compliance and interventions, such as facility inspections, technical assistance, and introduction of a voluntary program. These data could also be used to test the effects on compliance of environmental programs initiated by business, such as ISO 14001, the international environmental management standard, and responsible care. EPA and states could use the results of such evaluations to strengthen enforcement strategies.

Compliance and enforcement data bases also contain valuable information about facilities' environmental releases. The IDEA database includes data on facility releases to air, land, and water. EPA's Office of Pollution Prevention and Toxics has recently created the Risk Screening Environmental Indicators Model, which allows researchers to calculate the risks associated with facility releases. The model estimates the toxicity of chemicals that facilities report to EPA's Toxics Release Inventory and models their human exposure. It covers virtually all facilities reporting the data since 1990.

The EPA's Office of Environmental Information, created in 1998 to fill gaps in health and environmental data, has helped to make these data available, develop common standards for data from different geographic locations and environmental media, and ensure data accuracy. EPA assigns every facility required to report release information an identification number so that information can be integrated and accessed. Similar initiatives to improve the availability and utility of facility performance data have taken root at the state level. For example, the Environmental Compliance Consortium, a voluntary collaboration among state environmental agencies, uses publicly available performance data to identify strong programs and showcase them as national models.

The availability of such data on facility environmental releases makes it possible to compare the environmental performance of facilities that do and do not participate in specific policy interventions and to explore other determinants of firms' environmental performance, such as management commitment to environmental excellence or participation in EPA's National Environmental Performance Track, a program for "top" environmental performers. Understanding the effects of such organizational variables can help explain variation among facilities and suggest more effective policy approaches.

Private industry has recently been developing analytic tools for evaluating the performance of its own policies and programs. For example, in the late 1990s the International Organization for Standardization published ISO 14030, the environmental performance evaluation standard, a tool managers can use to compare their actual environmental performance to what they intended. Many individual firms devote substantial resources to evaluating their own environmental performance and that of their peers (see, e.g., http://www.intel.com/intel/other/ehs/perform.htm). Research could help strengthen these efforts by documenting and comparing the aspects of performance companies are evaluating.

Much more work still needs to be done to develop reliable and comprehensive outcome metrics, as noted in the discussion of environmental indicators above. For example, available data still tell little about the impact of policies geared to improving land use management, natural resource utilization, or pollutant emissions from nonpoint sources. They do not summarize consumption of energy, water, or other environmental inputs. Nonmanufacturing firms may not be required to report releases to government at all, even though their impacts may be significant. Improved data in such areas would strengthen researchers' capabilities to do sound evaluations.

Environmental policy evaluation can make a difference in policy by distinguishing promising policy approaches from those that are unlikely to lead to substantial advances. It can lead to adjustments in policy emphasis to achieve desired objectives, and improve decision makers' ability to create, compare, and implement more attractive policy alternatives.

Such research has a strong likelihood of being used because of increased emphasis on results-oriented government, especially in the federal government. Following the Government Performance and Results Act of 1993, a Performance Assessment Rating Tool (PART) was developed by the Office of Management and Budget "to systematically and routinely assess program performance" (Johnson, 2003a). Using PART, agencies are specifically asked whether they regularly conduct independent evaluations of their programs (Johnson, 2003b). In August 2001, President George W. Bush proclaimed results-oriented government as one of three overriding principles (U.S. Office of Management and Budget, 2002). Evaluation is central to the implementation of a results-oriented approach to government.

High-level, nonpartisan groups, such as Enterprise for the Environment and the National Academy of Public Administration, have called for results-based and information-driven environmental policy (Ruckelshaus and Hausker, 1998; National Academy of Public Administration, 1995, 1997). EPA's strategic plan for 2003-2008 lists focusing on results as a primary objective and its innovation strategy calls on environmental agencies to "emphasize results more than the means to achieve them, using regulatory and non-regulatory tools" (U.S. Environmental Protection Agency, 2002b). EPA has pledged to "evaluat[e] innovations results to make strategic decisions about those that can and should be applied on a broader scale" (U.S. Environmental Protection Agency, 2002b). This strategy acknowledges the critical role evaluation plays in determining which innovations are ready to be scaled up to full-blown programs, which require refinement, and which are unworkable.

Notwithstanding the demand for evaluation, it will be important for research to explore ways to design program evaluations so as to ensure they will be used. The barriers to use are fairly well documented (Solomon, 1998; Kraft, 1998). When evaluators do not understand the political contexts of the programs they study, they are apt to produce results that are not relevant to policy makers' questions. Evaluators may present results in complex and technical language or may fail to disseminate their findings widely. The structure of environmental agencies—organized by environmental media, staffed by regulatory experts, and reliant on outside contractors—may make them particularly resistant to utilizing evaluations. These barriers to use are not insurmountable, however.

In summary, environmental program evaluation is critical for improving the effectiveness of public and private environmental policies, and the need for measures of policy results has been recognized at the highest levels of the federal government. In some areas, particularly pollutant emissions from the manufacturing and utility sectors, useful data are now available on compliance and environmental releases. Evaluation is particularly salient in the context of widespread claims that traditional regulatory approaches to environmental protection are reaching the limits of their effectiveness. Many innovative approaches at the federal, state, and local levels have now been in use long enough to have established substantial track records. An ambitious agenda of environmental policy evaluation research could help fill the need for better understanding of what has worked and what has not and can thus provide better decision-relevant information for policy uses.

IMPROVING ENVIRONMENTAL FORECASTING

Federal environmental agencies should undertake an assortment of research initiatives to collect, appraise, develop, and extend analytic activities related to forecasting in order to improve environmental understanding and decision making.

Forecasting refers to a diverse collection of tools, methods, and practical approaches all generally intended to clarify past trends by their extension into the future, as well as to reveal and explore possibilities by postulating likely and desired changes in the present and projecting their consequences. Both forecasting and prediction aim to the future, creating a potential confusion between the terms. Forecasts may be assessed for their accuracy as predictions, but they should also be appraised and valued for their ability to reveal and discover phenomena, relationships, and the implications of existing trends. None of these important purposes operates in the same way when prediction is the goal (Brewer, 1992).

Many reasons exist to motivate forecasting activities. Among them are the desires to increase the available lead time until decisions must be made to allow more careful analysis of various options and associated outcomes and to increase the chance for broad public participation in decision making. Short lead times make for poorly considered decisions and limit potential participation (Anderson, 1997). To illuminate and secure the common interest and achieve the widest possible shaping and sharing of human values, time is needed to define and analyze problems, synthesize and communicate diverse and complex information, and weigh the legitimate interests of numerous stakeholders and participants. Forecasting, in its most general sense, has the potential to contribute significantly to all of these objectives.

The Research Need

Environmental forecasting, like the creation of indicators, has often suffered from inadequate appreciation of the relevant social science. For example, forecasting tools are often developed without taking into account the diverse needs of forecast users for information depending on their decision situations. Researchers often "do all the science" first, usually in search of some point prediction of likely events, and then "add on" the human dimensions almost as an afterthought. Consequently, the resulting forecasts often fail to connect to the needs of decision makers (Sarewitz and Pielke, 1999).

In addition, forecasts are often treated mechanistically. Climate change forecasting, for example, has predominantly been based on models of how greenhouse gases, once emitted, propagate through the global environment and bring about changes in global mean temperature, glacier thickness, and other physical variables. A mechanistic approach is appropriate so long as the forecasting task is limited to physical processes governed by universal laws. It has the advantage of being readily defensible as rational and objective (Ascher, 1987). Because the predictions are thought to be objective, controversy over policy is believed to be confined to settling differences in normative judgments (Friedman, 1953).

This approach has serious limitations when used to inform the decisions of human beings engaged in realistic policy decision making (Hammond, 1996). Clear testimony of its limits comes in the consistently poor predictive accuracy of forecasts using complex, computer-based economic and behavioral models (Ascher, 1978; Greenberger, Brewer, Hogan, and Russell, 1983; Craig, Gadgil, and Coomey, 2002; also see Appendix E). Although such models often have heuristic value, there has been no improvement since the early 1950s in the predictive accuracy of forecasts based on such models, which perform no better than judgmental forecasts. Complex models are less satisfactory than judgmental or scenario-based forecasts in terms of transparency, because they are commonly fitted to data by adjusting model assumptions and specifications. Among the beneficial roles of simple models is their capacity to explore alternative assumptions in a relatively straightforward and transparent manner—tasks quite different from simply predicting.

Forecasting efforts could be improved by focusing from the start on the human setting of environmental decision making, which should be the starting point and the framework within which past trends and possible futures are forecast. Forecasting efforts should encompass human influences on the environment as well as biophysical processes. They should be directed at decision-relevant outcomes, including environmental, health, and socioeconomic outcomes and the distribution of these outcomes across segments of the population. As with other aspects of environmental measurement, the development of forecasting methods should be guided by input from the potential users.

A human-centered approach stresses the concepts of intentionality and choice. Because people can invent and alter the future (sometimes influenced by forecasts), a simple "null projection" of past trends provides only a first-order approximation of future events or problems—one possible future out of many. The idea of "null" conveys the sense of what might happen if nothing changes—the status quo conditions of rules and relationships thrust into the future. Numerous plausible decisions need to be considered and analyzed to see which among them offer advantages compared with the null projection in terms of particular sets of values (Bobbitt, 2003; Hawken et al., 1982).

Forecasting, like the development of environmental indicators, presents substantial challenges in terms of acknowledging and consulting a diversity of information sources. These include applying the appropriate analytic tools to assemble, organize, integrate, and synthesize the available information into a useful whole and the need to present the results in meaningful and intelligible forms that can be readily and reliably communicated to potential users.

Environmental forecasting can benefit by drawing on experience with related kinds of forecasting (Armstrong, 2001). For example, models, simulations, games, and other closely related tools and methods have long been commonplace in the national security realm, but there has been little transfer or crossover of these tools to environmental forecasting. Efforts to conduct integrated environmental assessments are somewhat encouraging (Brewer, 1986), although comparisons of them to comparable military systems analyses show ample room for improvement. Among other things, because environmental assessments and forecasts typically offer a single-minded vision of the future, likely changes and plausible surprises and the consequent need for forecasting to be creative and adaptive are given short shrift. What is needed are multiple scenarios based on the perspectives and decision needs of many different participants and stakeholders that allow the exploration of many possible futures and conditions. The so-called Total Systems Performance Assessment (TSPA) for the proposed nuclear waste repository at Yucca Mountain, Nevada, is illustrative of the problem and this need (U.S. Nuclear Waste Technical Review Board, 1995).

Sophisticated private-sector environmental forecasting, such as in the energy area, may have limited social or public utility for proprietary reasons (Schwartz, 1996). The experiences of Stanford University's Energy Modeling Forum (an activity funded in large part by the Electric Power Research Institute) provide one strong example of positive developments from the private sector that can offer insights for public-sector environmental forecasting (Weyant et al., 1996).

Research Areas

Identifying Best Practices

We recommend a wide-ranging stock taking and appraisal of forecasting tools, methods, and experiences from a variety of different fields and subject matters to identify general best practices and to highlight common problems—the former to serve as exemplars and the latter to be avoided. Such an effort would go a long way toward enriching and improving forecasting in general practice and forecasts in the specific environmental realm. Taking stock and then setting and enforcing standards are desirable outcomes. Creating courses and curricula to emphasize best practices will contribute to improvements with longer term and longer lasting effects. As essential as forecast methods and tools are for integration and synthesis, it is remarkable to realize that they are not regularly taught in graduate educational or professional training programs.

Environmental Modeling Forums

To open access and to make forecasting more transparent and thus credible, we recommend sponsorship of one or a few continuing environmental modeling forums patterned on the long-running and successful Energy Modeling Forum at Stanford University. Providing ongoing and continuing appraisal is a desirable outcome here. The longerterm effects might include increased numbers of qualified professionals and improved understanding of environmental forecasts by government officials and the publics they serve. Enlarging the circle of those who understand the strengths and limitations of environmental models and analyses may also help to raise the credibility of forecasts and forecasters from their current low levels.

Improving Characterization of Uncertainty

At the heart of any forecast are difficult matters related to uncertainties. Uncertainties appear in the relevance and reliability of data, in the appropriateness of theories used to structure analyses, in the completeness of the specification of the problem, and in the "fit" between a forecast and the social and political matters of fact on the ground. Moreover, the characterization of uncertainty should consider the decision relevance of different aspects of the uncertainties. Failure to appreciate such uncertainties results in poor decisions, misinterpretation of forecasts, and to diminished trust of analysts among the potential users of forecasts. Considerable past work on uncertainty in environmental assessments and models make this topic ripe for progress (e.g., Morgan and Henrion, 1990; Rotman and van Asselt, 2001; National Research Council, 1997a). Improved ways to describe uncertainties in forecasts would provide widespread benefits.

Rationale for the Activity

The demand for environmental forecasts from public policy makers is evident in their continued reference to expected futures as providing the rationale for policy. The question is whether this field is ready to make significant scientific progress. The modest research efforts recommended here are highly leveraged because of past efforts in other problem areas and policy arenas and are likely to yield cost-effective results in terms of establishing best practices and standards for including and highlighting essential human aspects to be addressed in forecasting efforts. New technical and professional courses of study are an expected outcome that would increase the number of competent environmental analysts.

Another useful product could well be independent appraisal and certification of environmental forecasts, analyses, and analysts. Such are likely goals for the recommended Environmental Modeling Forums. Certification processes are likely to create incentives for improved clarity and transparency of forecast models and, in time, improve the low levels of trust and confidence the public generally has for environmental forecasts and those who use them.

DETERMINING DISTRIBUTIONAL IMPACTS

Federal agencies should support concerted efforts to improve the data, methods, and analytical techniques for determining the distributional impacts of environmental policies and programs related to issues of environmental inequities and their abatement. These efforts should include the determination of the most appropriate levels of social, spatial, and temporal aggregation of measurement for environmental monitoring and indicator development.

The Research Need

Concern about the distributional impacts of environmental risks has a long history in the United States and in EPA. Continuing concerns and controversy about claims of environmental inequity and injustice underline the importance of developing an adequate evidence base for addressing the issues.

Empirical research on these issues has been fragmented, inconclusive, and inconsistent in its results. The pioneering empirical work on distributional impacts in the 1970s focused on pollution in cities (Kruvant, 1974; Berry, 1977). Later, interest was driven by claims of environmental injustices based on documentation of the disproportionate burden of toxic waste on minority communities (U.S. General Accounting Office, 1983; United Church of Christ, 1987). Much of the recent literature on environmental inequity and injustice consists of activism and advocacy, analysis of the legal and civil rights aspects of environmental justice, and theoretical discussions of the meaning of equity (Bowen, 2001, 2002; English, 2004; Liu, 2001; Szasz and Meuser, 1997).

In recent years, there has been a marked increase in the number of methodologically based studies, especially those employing spatial analytical techniques (Stockwell, Sorenson, Eckert, and Carreras, 1993; Chakraborty and Armstrong, 1997; McMaster, Leitner, and Sheppard,

1997; Cutter et al., 2002; Mennis, 2002; Pine, Marx, and Lakshmanan, 2002) or historical demographic methods for measuring the evolution of inequities (Oakes, Anderton, and Anderson, 1996; Yandle and Burton, 1996; Been and Gupta, 1997; Mitchell, Thomas, and Cutter, 1999). Nevertheless, the measurement and modeling of environmental inequality and its causes are still in their infancy. There are also fundamental questions regarding the appropriate social and geographic scales for analyzing claims of inequity (Greenberg, 1993; Zimmerman, 1994; Cutter, Holm, and Clark, 1996) and the relationship of environmental justice issues to other issues of public policy decision making (Bowen and Wells, 2002; Sexton and Adgate, 1999; Margai, 2001; Miranda, Dolinoy, and Overstreet, 2002; Sexton, Waller, McMaster, Maldonado, and Adgate, 2002). Despite considerable research and policy interest during the past 20 years, fundamental questions remain concerning how to determine that environmental justice problems exist and, once determined, how to abate them. Four research themes seem most promising for addressing such questions.

Areas of Research

Defining Key Variables

Executive Order 12898 requires each federal agency to "make achieving environmental justice part of its mission," defines that mission with reference to adverse effects on "minority populations and low-incom ity can make a significant contribution to defining and measuring these concepts, taking into account changes in definitions over time and across space (e.g., the contextual nature of the terms).

Social scientists can also contribute to a deeper understanding of environmental justice in other ways. They can analyze whether the most significant adverse impacts are best identified by analyzing only residential location, as is often done, or by also considering occupational categories, prior health condition, or combinations of these and other risk factors. They can also help understand cultural, socioeconomic, or other systemic differences in what people or communities see as unjust and the conditions under which individuals and communities judge that an environmental injustice has been done.

Analyzing the Dependence of Impacts on Spatial and Temporal Scale

Associations between human activities and environmental conditions are well known to appear different as a function of the scale of measurement (e.g., Wilbanks and Kates, 1999; Geist and Lambin, 2002; Association of American Geographers Global Change in Local Places Working

Group, 2003). The implications of this general finding for environmental justice issues remain unknown. At present, data and analytical methods are inadequate to examine equity problems and impacts at multiple scales (e.g., individuals to ecosystems to regions). The existing research base is restricted geographically (focused on a few individual cities with census tracts as the enumeration unit) and temporally, usually providing only a static view of current distributions of impacts. There is little comparative work between urban places or between cities, suburbs, and rural areas. In urban settings, different settlement histories in northeastern versus southern or western cities raise questions about whether it is more appropriate to look for distributional impacts in central cities or in larger scale entities such as metropolitan areas or watersheds. Shifts in impacts over time have not yet been the subject of much investigation, yet they are extremely important for forecasting the future outcomes of environmental policies and programs and addressing issues of generational equity.

Developing Integrated Biophysical and Social Models

Models relevant to assessing differential exposures to and impacts of environmental risks are being developed by specialists in various fields—environmental, natural, and health sciences, social sciences, and engineering sciences—without much dialogue among researchers or their models. A sustained effort to model environmental impacts in an integrated way, focusing on their distribution, can help instigate this dialogue and lead to significant scientific advances. Integrated models should include (1) multiple stressors (cumulative or simultaneous); (2) multiple pathways of exposure (air, water, land); and (3) social vulnerability metrics. Consequently, efforts to build the models will engage a broad cross-section of researchers with critical issues in developing indicators, improving the quantity and quality of georeferenced data, and considering issues of scale dependence.

Improving Visualization and Risk Communication

Many distributional impacts and equity considerations are inherently geographical and readily displayed using maps. However, more interactive and sophisticated approaches to visualization, for example, using animations and virtual reality, may be very useful in this application arena because they may be accessible to a wider range of nonexpert audiences and they may enhance decision makers' understanding of the differential impact of risks and the variability in impacts from their management. The use of advanced geographic decision support tools may also result in a better informed next generation of American citizens. Examples of such tools now in use include the EPA's Surf Your Watershed (see http://www.epa.gov/surf/)

and Environmental Defense's Scorecard see http://www.scorecard.org). Such tools need evaluation, of course, in terms of their ability to provide accurate information about the distribution of environmental impacts, their value in generating ideas about how to reduce inequities, and their accessibility to the various populations concerned with environmental justice issues.

Rationale for the Activity

A concerted effort to measure the distribution of environmental impacts would advance both social science and environmental decision making. The research community is now mature enough to move from the rhetoric of environmental justice to the scientific analysis of the underlying phenomena. The proposed research activities can significantly advance understanding by scientists and the public of the dimensions and underlying causes of situations judged as inequitable through developing improved methods of measuring and monitoring them and models for understanding them. This would represent a significant advance over the evidence base for policy at present, which often consists of perceptions of environmental inequities among stakeholder groups rather than any robust empirically based assessment of exposures and impacts.

The recommended research would produce results that are potentially useful to policy makers at various levels of government and to citizens concerned with environmental justice issues. Moreover, there appear to be ready users. An EPA advisory panel has recommended that the agency make environmental justice a core part of its policies and expand its risk framework for measuring the cumulative impacts of toxic chemicals on disadvantaged communities (*Risk Policy Report*, March 16, 2004:8). This research would provide a tool kit to environmental agencies for examining the impacts of federal and state policies on particular localities or affected groups and considering whether or not the policies are achieving their desired results. The tools developed could help monitor progress toward environmental goals, such as embodied in Executive Order 12898, and ultimately reduce the unanticipated consequences of environmental decision making at the federal and state levels.

References

Acheson, J.M.

2003 *Capturing the Commons: Devising Institutions to Manage the Maine Lobster Industry*. Hanover, NH: University Press of New England.

Ackerman, B., and R.B. Stewart

1985 Reforming environmental law. *Stanford Law Review* 37:1333.

Ahmad, Y.J., S. El Serafy, and E. Lutz

1989 *Environmental Accounting for Sustainable Development*. Washington, DC: World Bank.

Allenby, B.R.

1992 Design for Environment: Implementing Industrial Ecology. Ph.D. Dissertation, Rutgers University.

1999 *Industrial Ecology: Policy Framework and Implementation*. Upper Saddle River, NJ: Prentice-Hall.

Anderson, D.D.

1997 Key concepts in anticipatory issues management. *Corporate Environmental Strategy* 5(1-Autumn):6-17.

Andrews, R.N.L.

1998 Environmental regulation and business "self-regulation." *Policy Sciences* 31(3):177-197.

1999 *Managing the Environment, Managing Ourselves: A History of American Environmental Policy*. New Haven, CT: Yale University Press.

Andrews, R.N.L., N. Darnall, D.R. Gallagher, S.T. Keiner, E. Feldman, M.L. Mitchell, D. Amaral, and J. Jacoby

2001 Environmental management systems: history, theory, and implementation research. Chapter 2 (pp. 31-60) in *Regulating from the Inside: Can Environmental Management Systems Achieve Policy Goals?* C. Coglianese and J. Nash, eds. Washington, DC: Resources for the Future Press.

Andrews, R.N.L., A.M. Hutson, and D. Edwards, Jr.
2004 Environmental management under pressure: how do mandates affect performance? In *Leveraging the Private Sector: Management Strategies for Environmental Performance*, C. Coglianese and J. Nash, eds. Washington, DC: Resources for the Future Press.

Aoyagi-Usui, M., H. Vinken, and A. Kuribayashi
2003 Pro-environmental attitudes and behaviors: An international comparison. *Human Ecology Review* 10:23-31.

Aragon-Correa, J.A., and S. Sharma
2003 A contingent resource-based view of proactive corporate environmental strategy. *Academy of Management Review* January:71-88.

Armstrong, J.S., ed.
2001 *Principles of Forecasting: A Handbook for Researchers and Practitioners*. Dordrecht, The Netherlands: Kluwer.

Arrow, K.
1963 *Social Choice and Individual Values*. New Haven, CT: Yale University Press.

Ascher, W.
1978 *Forecasting: An Appraisal for Policymakers and Planners*. Baltimore, MD: Johns Hopkins University Press.
1987 Subjectivity and the policy sciences. *Operant Subjectivity* 10(3):73-80.
2004 Scientific information and uncertainty: Challenges for the use of science in policymaking. *Science and Engineering Ethics* 10:437-455.

Ashford, N., and K. Rest
1999 *Public Participation in Contaminated Communities*. Cambridge, MA: Center for Technology, Policy, and Industrial Development, Massachusetts Institute of Technology.

Ashford, N., C. Ayers, and R.F. Stone
1985 Using regulation to change the market for innovation. *Harvard Environmental Law Review* 9:419.

Association of American Geographers Global Change in Local Places (GCLP) Working Group
2003 *Global Change and Local Places: Estimating, Understanding, and Reducing Greenhouse Gases*. Cambridge, England: Cambridge University Press.

Ayres, I., and J. Braithwaite
1992 *Responsive Regulation: Transcending the Deregulation Debate*. New York: Oxford University Press.

Bagozzi, R., Z. Gurhan-Canli, and J. Priester
2002 *The Social Psychology of Consumer Behavior*. Maidenhead, England: Open University Press.

Bamberg, S., and P. Schmidt
2003 Incentives, morality, or habit? Predicting students' car use for university routes with the models of Ajzen, Schwartz, and Triandis. *Environment and Behavior* 35:264-285.

Been, V., and F. Gupta
1997 Coming to the nuisance of going to the barrios? A longitudinal analysis of environmental justice claims. *Ecology Law Quarterly* 24(1):1-56.

Beierle, T., and J. Cayford
2002 *Democracy in Practice: Public Participation in Environmental Decisions*. Washington, DC: Resources for the Future Press.

Bell, D., H. Raiffa, and A. Tversky, eds.
1988 *Decision Making: Descriptive, Normative, and Prescriptive Interactions*. New York: Cambridge University Press.

Berkes, F.
2002 Cross-scale institutional linkages: Perspectives from the bottom up. Pp. 293-321 in National Research Council, *The Drama of the Commons*. Committee on the Human Dimensions of Global Change, E. Ostrom, T. Dietz, N. Dolsak, P.C. Stern, S. Stonich, and E.U. Weber, eds. Division of Behavioral and Social Sciences and Education. Washington, DC: National Academy Press.

Berry, B.J.L., ed.
1977 *The Social Burdens of Environmental Pollution: A Comparative Metropolitan Data Source*. Cambridge, MA: Ballinger.

Bijker, W.E., T.P. Hughes, and T. Pinch, eds.
1987 *The Social Construction of Technological Systems*. Cambridge, MA: MIT Press.

Bingham, G., and L.M. Langstaff
2003 *Alternative Dispute Resolution in the NEPA Process*. Washington, DC: Resolve, Inc.

Black, J.S., P.C. Stern, and J.T. Elworth
1985 Personal and contextual influences on household energy adaptations. *Journal of Applied Psychology* 70:3-21.

Boardman, A.E., D.H. Greenberg, A.R. Vining, and D.L. Weimer
2000 *Cost-Benefit Analysis: Concepts and Practice* (Second Edition). Upper Saddle River, NJ: Prentice Hall Business Publishing.

Bobbitt, P.
2003 Seeing the futures. *The New York Times,* Section A, p. 29, December 8.

Bommer, R.
1999 Environmental policy and industrial competitiveness: The pollution-haven hypothesis reconsidered. *Review of International Economics* 7(2):342-355.

Bowen, W.M.
2001 *Environmental Justice Through Research-Based Decision-Making*. New York: Garland Publishing.
2002 An analytic review of environmental justice research: What do we really know? *Environmental Management* 29:3-15.

Bowen, W.M., and M.V. Wells
2002 The politics and reality of environmental justice: A history and consideration for public administrators and policy makers. *Public Administration Review* 62(6): 688-698.

Bradbury, J.A., K.M. Branch, and E. Malone
2003 *An Evaluation of DOE-EM Public Participation Programs*. (Report prepared for the U.S. Department of Energy Environmental Management Program. #PNNL-14200.) Washington, DC: Pacific Northwest National Laboratory.

Brams, P., and P. Fishburn
2002 Voting procedures. Pp. 175-236 in *Handbook of Social Choice and Welfare,* K. Arrow, A. Sen, and K. Kuzumura, eds. Amsterdam, Holland: Elsevier.

Brewer, G.D.
1986 Methods for synthesis. Pp. 455-473 in *Sustainable Development of the Biosphere,* W.C. Clark and R.E. Munn, eds. Cambridge, England: Cambridge University Press.
1992 Discovery is not prediction. Chapter 6 in *Avoiding the Brink: Theory and Practice in Crisis Management,* A.C. Goldberg, D.V. Opstal, and J.H. Barkley, eds. London, England: Brassey's Defence Publications.

Brick, P., D. Snow, and S. Van de Wetering, eds.
2001 *Across the Great Divide: Explorations in Collaborative Conservation and the American West*. Washington, DC: Island Press.

Brower, M., and W. Leon
1999 *The Consumer's Guide to Effective Environmental Choices: Practical Advice from the Union of Concerned Scientists.* New York: Three Rivers Press.
Brown, B.B., C. Werner, and N. Kim
2003 Personal and contextual factors supporting the switch to transit use: Evaluating a natural transit intervention. *Analyses of Social Issues and Public Policy* 3:139-160.
Brown, M.A.
2001 Market failures and barriers as a basis for clean energy policies. *Energy Policy* 29:1197-1207.
Brown, M.A., M.D. Levine, and W. Short
2001 Scenarios for a clean energy future. *Energy Policy* 29:1179-1996.
Brunner, R.D., and W. Ascher
1992 Science and social responsibility. *Policy Sciences* 25:295-331.
Camerer, C., G. Loewenstein, and M. Rabin
2003 *Advances in Behavioral Economics.* Princeton, NJ: Princeton University Press.
Canadian Standards Association
1997 *Risk management guidelines for decision makers.* Ottawa, Ontario: Canadian Standards Association.
Caplan, N., A. Morrison, and R.J. Stambaugh
1975 *The Use of Social Science Knowledge in Policy Decisions at the National Level: A Report to Respondents.* Ann Arbor: University of Michigan.
Caro, F.G., ed.
1971 *Readings in Evaluation Research.* New York: Russell Sage Foundation.
Cash, D.W., and S.C. Moser
2000 Linking global and local scales: Designing dynamic assessment and management process. *Global Environmental Change* 10:109-120.
Chakraborty, J., and M. Armstrong
1997 Exploring the use of buffer analysis for the identification of impacted areas in environmental equity assessment. *Cartography and Geographic Information Systems* 24(3):145-157.
Chertow, M.R., and D.C. Esty, eds.
1997 *Thinking Ecologically: The Next Generation of Environmental Policy.* New Haven, CT: Yale University Press.
Chess, C.
1999 A model of organizational responsiveness to stakeholders. *Risk: Health, Safety and Environment* 10:257-267.
Chess, C., B.J. Hance, and G. Gibson
2000 Adaptive participation in watershed management. *Journal of Soil and Water Conservation Third Quarter* 2000:248-252.
Clemen, R.
1996 *Making Hard Decisions* (Second Edition). Belmont, CA: Duxbury Press.
Cohrssen, J., and V.T. Covello
1989 *Risk Analysis: A Guide to Principles and Methods for Analyzing Health and Environmental Risks.* Springfield, VA: National Technical Information Service.
Cole, D.H.
2002 *Pollution & Property: Comparing Ownership Institutions for Environmental Protection.* Cambridge, England: Cambridge University Press.
Cottrell, S.
2003 Influence of sociodemographics and environmental attitudes on general responsible environmental behavior among recreational boaters. *Environment and Behavior* 35:347-375.

Cowling, E.B.
1992 The performance and legacy of NAPAP. *Ecological Applications* 2(2):111-116.
Cowling, E.B., and J. Nilsson
1995 Acidification research: Lessons from history and visions of environmental futures. *Water, Air, and Soil Pollution* 85:279-292.
Craig, P.P., A. Gadgil, and J.G. Coomey
2002 What can history teach us? A retrospective examination of long-term energy forecasts for the United States. *Annual Review of Energy and Environment* 27:83-113.
Cronbach, L.J., S.R. Ambron, S.M. Dornbusch, R.D. Hess, R.C. Hornik, and D.C. Phillips
1980 *Toward a Reform of Program Evaluation.* San Francisco: Jossey-Bass.
Cropper, M.L., and W.E. Oates
1992 Environmental economics: A survey. *Journal of Economic Literature* 30:674-740.
Crouch, E., and R. Wilson
1982 *Risk/Benefit Analysis.* Cambridge, MA: Ballinger Publishing Company.
Cullen, A.C., and H.C. Frey
1999 *Probabilistic Techniques in Exposure Assessment.* New York: Plenum.
Cutter, S.L., D. Holm, and L. Clark
1996 The role of geographic scale in monitoring environmental justice. *Risk Analysis* 16(4):517-526.
Cutter, S.L., J.T. Mitchell, A. Hill, L. Harrington, S. Katkins, W. Muraco, J. DeHart, A. Reynolds, and R. Shudak
2002 Attitudes toward reducing greenhouse gas emissions from local places. Pp. 171-191 in *Global Change and Local Places: Estimating, Understanding, and Reducing Greenhouse Gases.* Association of American Geographers Global Change in Local Places (GCLP) Working Group. Cambridge, England: Cambridge University Press.
Dahlstrand, U., and A. Biel
1997 Pro-environmental habits: Propensity levels in behavioral change. *Journal of Applied Social Psychology* 27:588-601.
David, B., and J.C. Turner
1996 Studies in self-categorization and minority conversion: Is being a member of the out-group an advantage? *British Journal of Social Psychology* 35:179-199.
David, C., ed.
1997 *Western Public Lands and Environmental Politics.* Boulder, CO: Westview Press.
deLeon, P.
1999 The stages approach to the policy process. In *Theories of the Policy Process*, P. Sabatier, ed. Boulder, CO: Westview Press.
Deshpande, R.
1981 Action and enlightenment functions of research. *Knowledge: Creation, Diffusion, Utilization* 2:317-330.
Dietz, T., and P.C. Stern
1998 Science, values, and biodiversity. *BioScience* 48:441-444.
Dietz, T., E. Ostrom, and P.C. Stern
2003 The struggle to govern the commons. *Science* 302(Dec. 12):1907-1912.
Dowell, G., S. Hart, and B. Yeung
2000 Do corporate global environmental standards create or destroy market value? *Management Science* 46(8):1059-1074.
Druckman, D., B. Broome, and S. Korper
1988 Value differences and conflict resolution. *Journal of Conflict Resolution* 32(3): 489-510.

Dryzek, J.S.
1990 *Discursive Democracy: Politics, Policy, and Political Science.* Cambridge, England: Cambridge University Press.
2000 *Deliberative Democracy and Beyond: Liberals, Critics and Contestations.* Oxford, England: Oxford University Press.

Dunlap, R.E.
2002 An enduring concern: Light stays green for environmental protection. *Public Perspectives* September/October:10-14.

Dunlap, R.E., and R. Scarce
1991 The polls/poll trends: Environmental problems and protection. *Public Opinion Quarterly* 55:713-734.

DuPont, P.
2000 Communicating with whom? The effectiveness of appliance energy labels in the U.S. and Thailand. *Proceedings, American Council for an Energy-Efficient Economy* 8.39-8.54.

Durant, R., D. Fiorino, and R. O'Leary
2004 *Environmental Governance Reconsidered.* Cambridge, MA: MIT Press.

Durning, A.
1992 *How Much Is Enough? The Consumer Society and the Future of the Earth.* New York: W.W. Norton.

Egan, C., C.T. Payne, and J. Thorne
2000 Interim findings of an evaluation of the U.S. energy guide label. *Proceedings, American Council for an Energy-Efficient Economy* LBL-46061.

Elster, J., ed.
1998 *Deliberative Democracy.* Cambridge, England: Cambridge University Press.

English, M.R.
2004 Environmental risk and justice. Pp. 119-159 in *Risk Analysis and Society: An Interdisciplinary Characterization of the Field*, T. McDaniels and M.J. Small, eds. Cambridge, England: Cambridge University Press.

Erwin, D.H., and D.C. Krakauer
2004 Insights into innovation. *Science* 304(21 May 2004):1117-1119.

Ester, P., and R.A. Winett
1982 Toward more effective antecedent strategies for environmental programs. *Journal of Environmental Systems* 11:201-221.

Falk, A., E. Fehr, and U. Fischbacher
2002 Appropriating the commons: A theoretical explanation. Pp. 157-195 in National Research Council, *The Drama of the Commons.* Committee on the Human Dimensions of Global Change, E. Ostrom, T. Dietz, N. Dolsak, P.C. Stern, S. Stonich, and E.U. Weber, eds. Division of Behavioral and Social Sciences and Education. Washington, DC: National Academy Press.

Feder, B.J.
1999 Chemistry cleans up a factory. *The New York Times* Section 3, p. 1, July 18.

Feeny, D., F. Berkes, B.J. McCay, and J.M. Acheson
1990 Tragedy of the commons: Twenty-two years later. *Human Ecology* 18:1-19.

Fiorino, D.J.
1990 Citizen participation and environmental risk: A survey of institutional mechanisms. *Science, Technology, and Human Values* 15:226-243.
1996 Toward a new system of environmental regulation: The case for an industry sector approach. *Harvard Environmental Law Review* 26:457.

Fischer, F.
2000 *Citizens, Experts and the Environment: The Politics of Local Knowledge.* Durham, NC: Duke University Press.
Fischer, K., and J. Schott, eds.
1993 *Environmental Strategies for Industry: International Perspectives on Research Needs and Policy Implications.* Washington, DC: Island Press.
Fisher, R.J.
1997 *Interactive Conflict Resolution.* Syracuse, NY: Syracuse University Press.
Fishkin, J.S.
1991 *Democracy and Deliberation: New Directions for Democratic Reform.* New Haven, CT: Yale University Press.
Franklin Associates
1991 *Product Life-Cycle Assessment: Guidelines and Principles.* (EPA Report #68-CO-0003.) Washington, DC: U.S. Environmental Protection Agency.
Freeman, A.M., III
1993 *The Measurement of Environmental and Resource Values: Theory and Methods,* Second Edition. Washington, DC: Resources for the Future Press.
2003 Economics, incentives, and environmental policy. Pp. 201-221 in *Environmental Policy: New Directions for the Twenty-First Century,* Fifth Edition. N.J. Vig and M.E. Kraft, eds. Washington, DC: CQ Press.
Freudenburg, W.R.
1989 Social scientists' contributions to environmental management. *Journal of Social Issues* 45(1):133-152.
Freudenburg, W.R., and Gramling, R.
2002 Scientific expertise and natural resource decisions: Social science participation on interdisciplinary scientific committees. *Social Science Quarterly* 83:119-136.
Friedman, M.
1953 *Essays in Positive Economics.* Chicago, IL: University of Chicago Press.
Frosch, R., and N. Gallopoulos
1989 Strategies for manufacturing. *Scientific American* 144:144-152.
Funtowicz, S.O., and J.R. Ravetz
1992 Three types of risk assessment and the emergence of post-normal science. Pp. 251-274 in *Social Theories of Risk,* S. Krimsky and D. Golding, eds. Westport, CT: Praeger.
Garcia-Johnson, R.
2000 *Exporting Environmentalism: U.S. Multinational Chemical Corporations in Brazil and Mexico.* Cambridge, MA: MIT Press.
Gardner, G.T., and P.C. Stern
2002 *Environmental Problems and Human Behavior,* Second Edition. Boston, MA: Pearson Custom Publishers.
Gatersleben, B., L. Steg, and C. Vlek
2002 Measurement and determinants of environmentally significant consumer behavior. *Environment and Behavior* 34:335-362.
Geisler, C., and G. Daneker, eds.
2000 *Property and Values: Alternatives to Public and Private Ownership.* Washington, DC: Island Press.
Geist, H.J., and E.F. Lambin
2002 Proximate causes and underlying forces of tropical deforestation. *BioScience* 52(2):143-150.

Geller, E.S., R.A. Winett, and P.B. Everett
1982 *Preserving the Environment: New Strategies for Behavior Change*. New York: Pergamon.

Gereffi, G., J. Humphrey, and T. Sturgeon
2005 The governance of global value chains: An analytic framework. *Review of International Political Economy*, 12:78-104.

Gibson, C.C., M.A. McKean, and E. Ostrom, eds.
2000 *People and Forests: Communities, Institutions, and Governance*. Cambridge, MA: MIT Press.

Gormley, W., and D. Weimer
1999 *Organizational Report Cards*. Cambridge, MA: Harvard University Press.

Graedel, T.E., and B.R. Allenby
2003 *Industrial Ecology*, Second Edition. Upper Saddle River, NJ: Prentice-Hall.

Greenberg, D.H., and M.B. Mandell
1991 Research utilization in policymaking: A tale of two series (of social experiments). *Journal of Policy Analysis and Management* 10:633-656.

Greenberg, M.R.
1993 Proving environmental inequity in siting locally unwanted land uses. *Risk: Issues in Health & Safety* 4(Summer):235-252.

Greenberger, M., G.D. Brewer, W. Hogan, and M. Russell.
1983 *Caught Unawares: The Energy Decade in Retrospect*. Cambridge, MA: Ballinger.

Greer, L.E.
2000 Anatomy of a successful pollution reduction project: Large reductions in toxic wastes and emissions are achievable at many if not most company sites. *Environmental Science and Technology* 34(11):254A-261A.

Guagnano, G.T., P.C. Stern, and T. Dietz
1995 Influences on attitude-behavior relationships: A natural experiment with curbside recycling. *Environment and Behavior* 27:699-718.

Gunderson, L., C.S. Holling, and S. Light, eds.
1995 *Barriers and Bridges to the Renewal of Ecosystems and Institutions*. New York: Columbia University Press.

Gunningham, N., and P. Grabowsky
1998 *Smart Regulation: Designing Environmental Policy*. New York: Oxford University Press.

Haas, P.M.
2004a Addressing the global governance deficit. *Global Environmental Politics* 4(4):1-15.
2004b When does power listen to truth? A constructivist approach to the policy process. *Journal of European Public Policy* 11(4):569-592.

Hahn, R.W., and R.N. Stavins
1991 Incentive-based environmental regulation: A new era from an old idea? *Ecology Law Quarterly* 18:1.

Hammond, J., R. Keeney, and H. Raiffa
1999 *Smart Choices*. Cambridge, MA: Harvard Business School Press.

Hammond, K.R.
1996 *Human Judgment and Social Policy: Irreducible Uncertainty, Inevitable Error, Unavoidable Injustice*. New York: Oxford University Press.

Hann, C.M.
1998 Introduction: The embeddedness of property. Pp. 1-47 in *Property Relations: Renewing the Anthropological Tradition*, C.M. Hann, ed. Cambridge, England: Cambridge University Press.

Hanna, S.S., C. Folke, and K.G. Maeler, eds.
1996 *Rights to Nature: Ecological, Economic, Cultural, and Political Principles of Institutions for the Environment.* Washington, DC: Island Press.

Hardin, G.
1968 The tragedy of the commons. *Science* 162:1243-1248.

Harrison, K.
2002 Challenges in evaluating voluntary environmental programs. Pp. 263-282 in National Research Council, *New Tools for Environmental Protection: Education, Information, and Voluntary Measures*, Committee on the Human Dimensions of Global Change, T. Dietz and P.C. Stern, eds. Division of Behavioral and Social Sciences and Education. Washington, DC: National Academy Press.

Hart, S.L.
1995 A natural resource-based view of the firm. *Academy of Management Review* 20:986-1014.
1997 Beyond greening: Strategies for a sustainable world. *Harvard Business Review* 75(1): 66-76.

Hart, S.L., and G. Ahuja
1996 Does it pay to be green? *Business Strategy and the Environment* 5:30-37.

Hart, S.L., and M.B. Milstein
1999 Global sustainability and the creative destruction of industries. *Sloan Management Review* 41(1):23-33 (Fall).

Hastie, R., S.D. Penrod, and N. Pennington
1983 *Inside the Jury.* Cambridge, MA: Harvard University Press.

Haveman, R.H.
1970 *Public Expenditures and Policy Analysis.* Chicago, IL: Markham Publishing Co.

Hawken, P., J. Ogilvey, and P. Schwartz
1982 *Seven Tomorrows: The Potential Crises that Face Humankind and the Role of Choice in Determining the Future.* New York: Bantam.

Herb, J., S. Helms, and M.J. Jensen
2002 Harnessing the "power of information": Environmental right to know as a driver of sound environmental policy. Pp. 253-262 in National Research Council, *New Tools for Environmental Protection: Education, Information, and Voluntary Measures.* Committee on the Human Dimensions of Global Change, T. Dietz and P.C. Stern, eds. Division of Behavioral and Social Science and Education. Washington, DC: National Academy Press.

Hirst, E.
1987 *Cooperation and Community Conservation.* (Final report, Hood River Conservation Project, DOE/BP-11287-18.) Portland, OR: Pacific Power and Light and Bonneville Power Administration.

Hirst, E., L. Berry, and J. Soderstrom
1981 Review of utility home energy audit programs. *Energy* 6:621-630.

H. John Heinz Center for Science, Economics, and the Environment
1999 *Designing a Report on the State of the Nation's Ecosystems.* Washington, DC: H. John Heinz Center for Science, Economics, and the Environment.

Holling, C.S., ed.
1978 *Adaptive Environmental Impact Assessment and Management.* London, England: John Wiley.

Houck, O.
1994 Of bats, birds, and B-A-T: The convergent evolution of environmental law. *Mississippi Law Journal* 63:403.

Howard, R.
1989 Knowledge maps. *Management Science* 35(8):903-922.
Huberman, M.
1990 Linkage between researchers and practitioners: A qualitative study. *American Educational Research Journal* 27:363-391.
Humphrey, J., and H. Schmitz
2001 Governance in global value chains. *IDS Bulletin* 32(3):19-29.
Institute of Medicine
2003 *Dioxins and Dioxin-Like Compounds in the Food Supply: Strategies to Decrease Exposure*. Committee on the Implications of Dioxin in the Food Supply. Food and Nutrition Board. Washington, DC: The National Academies Press.
Irwin, A.
1995 *Citizen Science: A Study of People, Expertise and Sustainable Development*. London, England: Routledge.
Janis, I., and L. Mann
1977 *Decision Making: A Psychological Analysis of Conflict, Choice and Commitment*. New York: Free Press.
Johnson, C., III
2003a Statement of the Honorable Clay Johnson, III, Deputy Director for Management, Office of Management and Budget, before the Committee on Government Reform, U.S. House of Representatives, September 18, 2003. Available: http://www.whitehouse.gov/omb/legislative/testimony/cjohnson/030918_cjohnson.html [March 22, 2004].
2003b Memorandum for Heads of Departments and Agencies, Program Assessment Rating Tool (PART) Update. July 16, 2003. Available: http://www.whitehouse.gov/omb/memoranda/m03-17.pdf [March 22, 2004].
Jones, K., and H. Klein
1999 Lessons from 12 years of comparative risk projects. *Annual Review of Public Health* 20:159-172.
Kahneman, D., P. Slovic, and A. Tversky, eds.
1982 *Judgment Under Uncertainty: Heuristics and Biases*. New York: Cambridge University Press.
Kahneman, D., I. Ritov, K.E. Jacowitz, and P. Grant
1993 Stated willingness to pay for public goods: A psychological perspective. *Psychological Science* 4:310-315.
Kaiser, F.G.
1998 A general measure of ecological behavior. *Journal of Applied Social Psychology* 28:395-422.
Kaplan, M.F., and C.E. Miller
1987 Group decision making and normative versus informational influence: Effects of type of issue and assigned decision rule. *Journal of Personality and Social Psychology* 53:306-313.
Kasemir, B., J. Jäger, C. Jaeger, and M.T. Gardner, eds.
2003 *Public Participation in Sustainability Science: A Handbook*. Cambridge, England: Cambridge University Press.
Kats, G.
2003 *The Costs and Financial Benefits of Green Buildings. Report to California's Sustainable Building Task Force*. Available: http://www.usgbc.org/Docs/News/News477.pdf [March 29, 2004].

Katzev, R.
2003 Car sharing: A new approach to urban transportation problems. *Analyses of Social Issues and Public Policy* 3:65-86.

Keeney, R.L.
1992 *Value-Focused Thinking: A Path to Creative Decisionmaking*. Cambridge, MA: Harvard University Press.

Keeney, R.L., and H. Raiffa
1993 *Decisions with Multiple Objectives,* Second Edition. New York: Cambridge University Press.

Kelleher, G., and C. Reccia
1998 Editorial—Lessons from marine protected areas around the world. *Parks* 8:1-4.

Kelman, S.
1981 Cost-benefit analysis: An ethical critique. *Regulation* 5(1):33-40.

Kempton, W., C.K. Harris, J.G. Keith, and J.S. Weihl
1985 Do consumers know "what works" in energy conservation? *Marriage and Family Review* 9:115-133.

Kingdon, J.
1987 *Agendas, Alternatives and Public Policies*. Boston, MA: Addison Wesley.

Kirschner, E.
1994 Full-cost accounting for the environment. *Chemical Week* 154(9):25-26.

Klassen, R D., and C.P. McLaughlin
1996 The impact of environmental management on firm performance. *Management Science* 72(3):2-7.

Klassen, R.D., and D.C. Whybark
1999 The impact of environmental technologies on manufacturing performance. *Academy of Management Journal* 42(6):599-615.

Kleindorfer, P.R., H.C. Kunreuther, and P.J.H. Shoemaker
1993 *Decision Sciences: An Integrative Perspective*. New York: Cambridge University Press.

Knapp, G.J., and T.J. Kim, eds.
1998 *Environmental Program Evaluation: A Primer*. Chicago: University of Illinois Press.

Knaap, G.J., and J.K. Tschangho, eds.
1998 *Environmental Program Evaluation: A Primer*. Chicago: University of Illinois Press.

Knight, R.L., and S.F. Bates, eds.
1995 *A New Century for Natural Resources Management*. Washington, DC: Island Press.

Kollman, K., and A. Prakash
2002 EMS-based environmental regimes as club goods: Examining variations in firms-level adoption of ISO 14001 and EMAS in UK, US, and Germany. *Policy Sciences* 35(1):43-67.

Kopelman, S.J., M. Weber, and D.M. Messick
2002 Factors influencing cooperation in commons dilemmas: A review of experimental psychological research. Pp. 113-156 in National Research Council, *The Drama of the Commons*. Committee on the Human Dimensions of Global Change, E. Ostrom, T. Dietz, N. Dolsak, P.C. Stern, S. Stonich, and E.U. Weber, eds. Division of Behavioral and Social Sciences and Education. Washington, DC: National Academy Press.

Kraft, M.E.
1998 Using environmental program evaluation: Politics, knowledge, and policy change. In *Environmental Program Evaluation: A Primer*, G. Knaap, J. Gerrit, and J.K. Tschangho, eds. Chicago: University of Illinois Press.

Krimsky, S., and D. Golding, eds.
1992 *Social Theories of Risk*. Westport, CT: Praeger.
Kruvant, W.
1974 *Incidence of Pollution Where People Live in Washington*. Washington, DC: Center for Metropolitan Studies.
Landry, R., N. Amara, and M. Lamari
2001 Utilization of social science research knowledge in Canada. *Research Policy* 30: 333-349.
Landry, R., M. Lamari, and N. Amara
2003 The extent and determinants of the utilization of university research in government agencies. *Public Administration Review* 63:192-205.
Langbein, L.
2002 Responsive bureaus, equity, and regulatory negotiation: An empirical view. *Journal of Policy Analysis and Management* 21:449-465.
Lave, L.B.
1996 Benefit-cost analysis: Do the benefits exceed the costs? Pp. 104-134 in *Risks, Costs, and Lives Saved: Getting Better Results from Regulation*, R. Hahn, ed. London, England: Oxford University Press.
Leach, W.D., N.W. Pelkey, and P.A. Sabatier
2002 Stakeholder partnerships as collaborative policymaking: Evaluation criteria applied to watershed management in California and Washington. *Journal of Policy Analysis and Management* 21:645-670.
Lee, J.
2004 E.P.A. may tighten its proposal on mercury. *New York Times*, March 16. Available: http://www.nytimes.com/2004/03/16/politics/16ENVI.html [June 1, 2004].
Lee, K.
1993 *Compass and Gyroscope: Integrating Science and Politics for the Environment*. Washington, DC: Island Press.
Leontief, W.W.
1966 *Input-Output Economics*. New York: Oxford University Press.
Levine, J.M., and R.L. Moreland
1998 Small groups. Pp. 415-469 in *The Handbook of Social Psychology*, Volume 2, Fourth Edition. D.T. Gilbert, S.T. Fiske, and G. Lindzey, eds. New York: McGraw-Hill.
Levy, D.L.
1995 The environmental practices and performance of transnational corporations. *Transnational Corporations* 4(1):44-67.
Lewis, H.W., R.J. Budnitz, H.J.C. Kouts, F. von Hippel, W. Lowenstein, and F. Zachariasen
1975 *Risk Assessment Group Report to the U.S. Nuclear Regulatory Commission*. Washington, DC: U.S. Nuclear Regulatory Commission.
Lindblom, C.E.
1959 The science of "muddling through." *Public Administration Review* 19(Spring): 79-88.
Lindblom, C.E., and D.K Cohen
1979 *Usable Knowledge: Social Science and Social Problem Solving*. New Haven, CT: Yale University Press.
Li, S.
2003 Recycling behavior under China's social and economic transition: The case of metropolitan Wuhan. *Environment and Behavior* 35:784-801.

Liu, F.
2001 *Environmental Justice Analysis: Theories, Methods, and Practice.* New York: Lewis Publishers.

Lowry, W.R.
2003 A new era in natural resource policies? Pp. 325-345 in *Environmental Policy: New Directions in the Twenty-First Century,* Fifth Edition, N.J. Vig and M.E. Kraft, eds. Washington, DC: CQ Press.

Lowy, D.C., and R.P. Wells
2000 *Corporate Environmental Governance: Benchmarks Toward World-Class Systems.* New York: The Conference Board.

Lutzenhiser, L.
1997 Social structure, culture and technology: Modeling the driving forces of household consumption. Pp. 77-91 in National Research Council, *Environmentally Significant Consumption: Research Directions,* Committee on the Human Dimensions of Global Change, P.C. Stern, T. Dietz, V.W. Ruttan, R.H. Socolow, and J. Sweeney, eds. Commission on Behavioral and Social Sciences and Education. Washington, DC: National Academy Press.
2002 Greening the economy from the bottom-up? Lessons in consumption from the energy case. Pp. 345-356 in *Readings in Economic Sociology,* N.W. Biggart, ed. Oxford, England: Blackwell.

Lutzenhiser, L., C. Harris, and M. Olsen
2001 Energy, society and environment. Pp. 222-271 in *Handbook of Environmental Sociology,* R. Dunlap and W. Michaelson, eds. Westport, CT: Greenwood Press.

Lutzenhiser, L., R. Kunkle, J. Woods, and S. Lutzenhiser
2003 Conservation behavior by residential consumers before and after the 2000-2001 California energy crisis. Pp. 147-200 in *Integrated Energy Policy Report: Public Interest Energy Strategies.* (Report 100-03-012f.) Sacramento, CA: California Energy Commission.

March, J.
1978 Bounded rationality, ambiguity and the engineering of choice. *Bell Journal of Economics* 9(2):587-608.
1997 *A Primer on Decision Making.* New York: Basic Books.

Margai, F.L.
2001 Health risks and environmental inequity: A geographical analysis of accidental releases of hazardous materials. *Professional Geographer* 53(3):422-434.

Marine Fish Conservation Network
2004 *Individual Fishing Quotas: Environmental, Public Policy, and Socioeconomic Impacts.* Washington, DC: Marine Fish Conservation Network.

Mazurek, J.
2002 Government-sponsored voluntary programs for firms: An initial survey. Pp. 219-234 in National Research Council, *New Tools for Environmental Protection, Environmental Protection: Education, Information, and Voluntary Measures,* Committee on the Human Dimensions of Global Change, T. Dietz and P.C. Stern, eds. Division of Behavioral and Social Sciences and Education. Washington, DC: National Academy Press.

McCay, B.J.
2000 Property rights, the commons, and natural resource management. Pp. 67-82 in *Property Rights, Economics, and the Environment.* M.D. Kaplowitz, ed. Stamford, CT: JAI Press.

2002 Emergence of institutions for the commons: Contexts, situations, and events. Pp. 361-402 in National Research Council, *The Drama of the Commons*, Committee on the Human Dimensions of Global Change, E. Ostrom, T. Dietz, N. Dolsak, P.C. Stern, S. Stonich, and E.U. Weber, eds. Division of Behavioral and Social Sciences and Education. Washington, DC: National Academy Press.

2004 ITQs and community: An essay on environmental governance. *Review of Agricultural and Resource Economics* 33(2):162-170.

McCay, B.J., and J.M. Acheson

1987 *The Question of the Commons: The Culture and Ecology of Communal Resources*. Tucson: University of Arizona Press.

McCay, B.J., and S. Brandt

2001 Changes in fleet capacity and ownership of harvesting rights in the united states surf clam and ocean quahog fisher. Pp. 44-60 in *Case Studies on the Effects of Transferable Fishing Rights on Fleet Capacity and Concentration of Quota Ownership*. R. Shotton, ed. (FAO Fisheries Technical Paper 412.) Rome, Italy: Food and Agriculture Organization of the United Nations.

McComas, K.A.

2001 Theory and practice of public meetings. *Communication Theory* 11:36-55.

McDaniels, T., and K. Thomas

1999 Eliciting public preferences for local land use alternatives: A structured value referendum with approval voting. *Journal of Policy Analysis and Management* 18(2):264-280.

McKenzie-Mohr, D., and W. Smith

2000 *Fostering Sustainable Behavior: An Introduction to Community-Based Social Marketing*. Gabriola Island, BC, Canada: New Society Publishers.

McMaster, R.B., H. Leitner, and E. Sheppard

1997 GIS-based environmental equity and risk assessment: Methodological problems and prospects. *Cartography and Geographic Information Systems* 24:172-189.

Mendelberg, T.

2002 The deliberative citizen: Theory and evidence. Pp. 151-193 in *Research in Micropolitics Vol. 6: Political Decision Making, Deliberation and Participation*. M. Delli Carpini, L. Huddy, and R.Y. Shapiro, eds. St. Louis, MO: Elsevier Science & Technology Books.

Mennis, J.

2002 Using geographic information systems to create and analyze statistical surfaces of population and risk for environmental justice analysis. *Social Science Quarterly* 83:281-297.

Metzenbaum, S.

2003 More nutritious beans. *Environmental Forum* (March/April):18-41.

Mikula, G.

2003 Testing an attribution-of-blame model of judgments of injustice. *European Journal of Social Psychology* 33:793-811.

Mileti, D.S., and L.A. Peek

2002 Understanding individual and social characteristics in the promotion of household disaster preparedness. Pp. 125-139 in National Research Council, *New Tools for Environmental Protection: Education, Information, and Voluntary Measures*. Committee on the Human Dimensions of Global Change, T. Dietz and P.C. Stern, eds. Division of Behavioral and Social Sciences and Education. Washington, DC: National Academy Press.

Miller, D.T.
2001 Disrespect and the experience of injustice. *Annual Review of Psychology* 52: 527-553.

Miranda, M.L., D.C. Dolinoy, and M.A. Overstreet
2002 Mapping for prevention: GIS models for directing childhood lead poisoning prevention programs. *Environmental Heath Perspectives* 110(9):947-953.

Mitchell, J.T., D.S.K. Thomas, and S.L. Cutter
1999 Dumping in Dixie revisited: The evolution of environmental injustices in South Carolina. *Social Science Quarterly* 80(2):229-243.

Morgan, M.G., and M. Henrion
1990 *Uncertainty*. Cambridge, England: Cambridge University Press.

Morgan, M.G., M. Kandlikar, J. Risbey and H. Dowlatabadi
1999 Why conventional tools of policy analysis are often inadequate for problems of global change. *Climatic Change* 41:271-281.

Morgan, M.G., B. Fischhoff, A. Bostrom, and C. Atman
2002 *Risk Communication: A Mental Models Approach*. Cambridge, England: Cambridge University Press.

Morgenstern, R.D., ed.
1997 *Economic Analysis at the EPA*. Washington, DC: Resources for the Future Press.

Moscovici, S.
1985 Innovation and minority influence. Pp. 347-412 in *Handbook of Social Psychology*, Volume 2, G. Lindzey and E. Aronson, eds. New York: Random House.

Moss, R.H, and S.H. Schneider
2000 Uncertainties in the IPCC TAR: Recommendations to lead authors for more consistent assessment reporting. Pp. 33-51 in *Guidance Papers on the Cross Cutting Issues of the Third Assessment Report*, R. Pachauri, T. Taniguchi, and K. Tanaka, eds. Geneva, Switzerland: World Meteorological Organization.

National Academy of Public Administration
1995 *Setting Priorities, Getting Results*. Washington, DC: National Academy of Public Administration.
1997 *Resolving the Paradox of Environmental Protection*. Washington, DC: National Academy of Public Administration.

National Research Council
1984 *Energy Use: The Human Dimension*. Committee on Behavioral and Social Aspects of Energy Consumption and Production, P.C. Stern and E. Aronson, eds. New York: Freeman.
1989 *Improving Risk Communication*. Committee on Risk Perception and Communication. Washington, DC: National Academy Press.
1992 *Global Environmental Change: Understanding the Human Dimensions*. P.C. Stern, O.R. Young, and D. Druckman, eds. Committee on the Human Dimensions of Global Change, Commission on Behavioral and Social Sciences and Education. Washington, DC: National Academy Press.
1994a *Science and Judgment in Risk Assessment*. Committee on Risk Assessment of Hazardous Air Pollutants, Board on Environmental Studies and Toxicology. Commission on Life Sciences. Washington, DC: National Academy Press.
1994b *Science Priorities for the Human Dimensions of Global Change*. Committee on the Human Dimensions of Global Change. Commission on Behavioral and Social Sciences and Education. Washington, DC: National Academy Press.

1995 *Technical Bases for Yucca Mountain Standards.* Committee on Technical Bases for Yucca Mountain Standards, Board on Radioactive Waste Management. Commission on Geosciences, Environment, and Resources. Washington, DC: National Academy Press.

1996 *Understanding Risk: Informing Decisions in a Democratic Society.* P.C. Stern and H.V. Fineberg, eds. Committee on Risk Characterization. Commission on Behavioral and Social Sciences and Education. Washington, DC: National Academy Press.

1997a *Review of Recommendations for Probabilistic Seismic Hazard Analysis: Guidance on Uncertainty and Use of Experts.* Panel on Seismic Hazard Evaluation. Committee on Seismology. Commission on Geosciences, Environment, and Resources. Washington, DC: National Academy Press.

1997b *Environmentally Significant Consumption: Research Directions.* P.C. Stern, T. Dietz, V.R. Ruttan, R. Socolow, and J. Sweeney, eds. Committee on the Human Dimensions of Global Change. Commission on Behavioral and Social Sciences and Education. Washington, DC: National Academy Press.

1999a *Downstream: Adaptive Management of Glen Canyon Dam and the Colorado River Ecosystem.* Committee on Grand Canyon Monitoring and Research. Water Science and Technology Board. Commission on Geosciences, Environment, and Resources. Washington, DC: National Academy Press.

1999b *Human Dimensions of Global Environmental Change: Research Pathways for the Next Decade.* Committee on the Human Dimensions of Global Change, Commission on Behavioral and Social Sciences and Education. Committee on Global Change Research, Board on Sustainable Development, Policy Division. Washington, DC: National Academy Press.

1999c *Nature's Numbers: Expanding the National Economic Accounts to Include the Environment.* W.D. Nordhaus and E.C. Kokkelenberg, eds. Panel on Integrated Environmental and Economic Accounting. Committee on National Statistics, Commission on Behavioral and Social Sciences and Education. Washington, DC: National Academy Press.

1999d *Making Climate Forecasts Matter.* P.C. Stern and W.E. Easterling, eds. Panel on the Human Dimensions of Seasonal-to-Interannual Climate Variability. Committee on the Human Dimensions of Global Change, Commission on Behavioral and Social Sciences and Education. Washington, DC: National Academy Press.

1999e *Perspectives on Biodiversity: Valuing Its Role in an Everchanging World.* Committee on Noneconomic and Economic Value of Biodiversity. Board on Biology, Commission on Life Sciences. Washington, DC: National Academy Press.

1999f *Materials Science and Engineering: Forging Stronger Links to Users.* National Materials Advisory Board, Commission on Engineering and Technical Systems. Washington, DC: National Academy Press.

2000 *Ecological Indicators for the Nation.* Committee to Evaluate Indicators for Monitoring Aquatic and Terrestrial Environments, Board on Environmental Studies and Technology. Water Science and Technology Board, Commission on Geosciences, Environment, and Resources. Washington, DC: National Academy Press.

2001a *Disposition of High-Level Waste and Spent Nuclear Fuel: The Continuing Societal and Technical Challenges.* Committee on Disposition of High-Level Radioactive Waste Through Geological Isolation, Board on Radioactive Waste Management. Board on Radioactive Waste Management, Division on Earth and Life Studies. Washington, DC: National Academy Press.

2001b *Grand Challenges in Environmental Sciences.* Committee on Grand Challenges in Environmental Sciences. Oversight Commission for the Committee on Grand Challenges in Environmental Sciences. Washington, DC: National Academy Press.

2002a *The Drama of the Commons.* Committee on the Human Dimensions of Global Change, E. Ostrom, T. Dietz, N. Dolsak, P.C. Stern, S. Stonich, and E.U. Weber, eds. Division of Behavioral and Social Sciences and Education. Washington, DC: National Academy Press.

2002b *New Tools for Environmental Protection: Education, Information, and Voluntary Measures.* Committee on the Human Dimensions of Global Change, T. Dietz and P.C. Stern, eds. Division of Behavioral and Social Sciences and Education. Washington, DC: National Academy Press.

2002c *Community and Quality of Life: Data Needs for Informed Decision Making.* Committee on Identifying Data Needs for Place-Based Decision Making, Committee on Geography. Board on Earth Sciences and Resources, Division on Earth and Life Studies. Washington, DC: National Academy Press.

2002d *Estimating the Public Health Benefits of Proposed Air Pollution Regulations.* Committee on Estimating the Health-Risk-Reduction Benefits of Proposed Air Pollution Regulations. Board on Environmental Studies and Toxicology. Washington, DC: The National Academies Press.

2003 *One Step at a Time: The Staged Development of Geologic Repositories for High-Level Radioactive Waste.* Committee on Principles and Operational Strategies for Staged Repository Systems. Board on Radioactive Waste Management. Division on Earth and Life Studies. Washington DC: The National Academies Press.

2004a *Materials Count: The Case for Material Flows Analysis.* Committee on Material Flows Accounting of Natural Resources, Products, and Residuals, Board on Earth Sciences and Resources. Division on Earth and Life Studies. Washington, DC: The National Academies Press.

2004b *Adaptive Management for Water Resources Project Planning.* Panel on Adaptive Management for Resource Stewardship. Committee to Assess the U.S. Army Corps of Engineers Methods of Analysis and Peer Review for Waste Resources Project Planning. Water Science and Technology Board. Ocean Studies Board. Division on Earth and Life Sciences. Washington, DC: The National Academies Press.

2004c *Implementing Climate and Global Change Research: A Review of the Final U.S. Climate Change Science Program Strategic Plan.* Committee to Review the U.S. Climate Change Science Program Strategic Plan. Division on Earth and Life Studies. Division of Behavioral and Social Sciences and Education. Division on Engineering and Physical Sciences. Washington, DC: The National Academies Press.

Norberg-Bohm, V.

2000 Beyond the double dividend: Public and private roles in the supply of and demand for environmentally enhancing technologies. Pp. 123-135 in Organisation for Economic Co-operation and Development, *Innovation and the Environment.* Paris, France: Organisation for Economic Co-operation and Development.

Nordlund, A.M., and J. Garvill

2002 Value structures behind proenvironmental behavior. *Environment and Behavior* 34:740-756.

Oakes, J.M., D.L. Anderton, and A.B. Anderson

1996 A longitudinal analysis of environmental equity in communities with hazardous waste facilities. *Social Science Research* 25(2):125-148.

Oh, C.H.

1996a Information searching in governmental bureaucracies: An integrated model. *American Review of Public Administration* 26:41-70.

1996b *Linking Social Science Information to Policy-Making.* W. Breit and K.G. Elzinga, eds. Political Economy and Public Policy Series. Greenwich, CT: JAI Press Inc.

Oh, C.H., and R.F. Rich
1996 Explaining use of information in pubic policy-making. *Knowledge and Policy: The International Journal of Knowledge Transfer and Utilization* 9(1):3-35.
Ölander, F., and J. Thøgerson
1995 Understanding consumer behavior as a prerequisite for environmental protection. *Journal of Consumer Policy* 18:345-385.
Olson, M.
1965 *The Logic of Collective Action.* Cambridge, MA: Harvard University Press.
Oom do Valle, P., E. Reis, J. Manazes, and E. Rebelo
2004 Behavioral determinants of household recycling participation: The Portuguese case. *Environment and Behavior* 36:505-540.
Organisation for Economic Co-operation and Development
2000 *Innovation and the Environment.* Paris, France: Organisation for Economic Co-operation and Development.
Orts, E.W.
1995 Reflexive environmental law. *Northwestern University Law Review* 89:1227.
Ostrom, E.
1990 *Governing the Commons: The Evolution of Institutions for Collective Action.* New York: Cambridge University Press.
2001 Decentralization and development: The new panacea. Pp. 237-256 in *Challenges to Democracy: Ideas, Involvement and Institution,* K. Dowding, J. Hughes, and H. Margetts, eds. New York: Palgrave Publishers.
Oversight Review Board of the National Acid Precipitation Assessment Program (M. Russell, K. Arrow, J. Bailar, J. Gordon, G. Hilst, S. Levine, T. Malone, W. Nierenberg, C. Starr, and J. Tukey)
1991 *The Experience and Legacy of NAPAP. Report to the Joint Chairs Council of the Interagency Task Force on Acidic Deposition.* Washington, DC: National Acid Precipitation Assessment.
Pashigian, B.P.
1985 Environmental regulation: Whose self-interests are being protected? *Economic Inquiry* 23(4):551-584.
Payne, J., J. Bettman, and E. Johnson
1993 *The Adaptive Decision Maker.* New York: Cambridge.
Pfirman, S., and the NSF Advisory Committee for Environmental Research and Education
2003 *Complex Environmental Systems: Synthesis for Earth, Life, and Society in the 21st Century: A 10-Year Outlook for the National Science Foundation.* Arlington, VA: National Science Foundation.
Pildes, R.H., and C.R. Sunstein
1995 Reinventing the regulatory state. *University of Chicago Law Review* 62 U:1.
Pine, J.C., B.D. Marx, and A. Lakshmanan
2002 An examination of accidental-release scenarios from chemical-processing sites: The relation of race to distance. *Social Science Quarterly* 83(1):317-331.
Poortinga, W., L. Steg, and C. Vlek
2004 Values, environmental concern, and environmental behavior: A study into energy use. *Environment and Behavior* 36:70-93.
Porter, D.R., and D.A. Salvesen, eds.
1995 *Collaborative Planning for Wetlands and Wildlife.* Washington, DC: Island Press.
Portney, P.
1990 *Public Policies for Environmental Protection.* Washington, DC: Resources for the Future Press.

Potoski, M., and A. Prakash

2005 Green Clubs and Voluntary Governance: ISO 14001 and Firms' Regulatory Compliance. *American Journal of Political Science* 49(2).

Powell, M.R.

1997 *Three-City Air Study*. (Discussion Paper 97-29.) Washington, DC: Resources for the Future Press.

1999 *Science at EPA: Information in the Regulatory Process*. Washington, DC: Resources for the Future Press.

Prakash, A.

2001 Why do firms adopt beyond-compliance environmental policies? *Business Strategy and the Environment* 10:286-299.

Presidential/Congressional Commission on Risk Assessment and Risk Management

1997 *Framework for Environmental Health Risk Management*. Volume 1. Washington, DC: Presidential/Congressional Commission on Risk Assessment and Risk Management.

Press, D.

1994 *Democratic Dilemmas: Trees and Toxics in the American West*. Durham, NC: Duke University Press.

Rabe, B.

2003 Power to the states: The promise and pitfalls of decentralization. Pp. 33-56 in *Environmental Policy: New Directions for the Twenty-First Century*, Fifth Edition, N.J. Vig and M.E. Kraft, eds. Washington, DC: CQ Press.

Raiffa, H.

1968 *Decision Analysis: Introductory Lectures on Choice Under Uncertainty*. Reading, MA: Addison-Wesley.

Redefining Progress

2004 *Calculate Your Ecological Footprint: 13 Simple Questions Will Assess Your Use of Nature*. Available: http://www.lead.org/leadnet/footprint/intro.htm [March 2005].

Rees, W., and M. Wackernagel

1995 *Our Ecological Footprint: Reducing Human Impacts on the Earth*. Gabriola, BC: New Society Publishers.

Renn, O.

2003 The challenge of integrating deliberation and expertise: Participation and discourse in risk management. Pp 289-366 in *Risk Analysis and Society: An Interdisciplinary Characterization of the Field*, T. McDaniels and M.J. Small, eds. Cambridge, England: Cambridge University Press.

Renn, O., T. Webler, and P. Wiedemann

1995 *Fairness and Competence in Citizen Participation: Evaluating New Models for Environmental Discourse*. Dordrecht, The Netherlands: Kluwer.

Ribot, J.C.

2002 *African Decentralization: Local Actors, Powers and Accountability*. (UNRIISD Programme on Democracy, Governance and Human Rights, Paper No. 8.) Geneva, Switzerland: United Nations Research Institute for Social Development.

Rich, R.F.

2001 *Social Science Information and Public Policy Making*, Second Edition. San Francisco, CA: Jossey-Bass.

Rodricks, J.V.

1992 *Calculated Risks: Understanding the Toxicity and Human Health Risks of Chemicals in our Environment*. Cambridge, England: Cambridge University Press.

Rose, C.M.
2002 Common property, regulatory property, and environmental protection: Comparing community-based management to tradable environmental allowances. Pp. 233-257 in National Research Council, *The Drama of the Commons.* Committee on the Human Dimensions of Global Change, E. Ostrom, T. Dietz, N. Dolsak, P.C. Stern, S. Stonich, and E.U. Weber, eds. Division of Behavioral and Social Sciences and Education. Washington, DC: National Academy Press.
Rossi, P.H., and H.E. Freeman
1993 *Evaluation: A Systematic Approach.* Newbury Park, CA: Sage Publications.
Rotman, J., and M. van Asselt
2001 Uncertainty management in integrated assessment modeling. *Environmental Monitoring and Assessment* 69:101-130.
Royal Commission on Environmental Protection
1998 *Setting Environmental Standards.* London, England: Royal Commission on Environmental Protection.
Royston, M.
1979 *Pollution Prevention Pays.* London, England: Pergamon Press.
1980 Making pollution prevention pay. *Harvard Business Review* 58(6):6-27
Rubin, J.Z., D.G. Pruitt, and S. Kim
1994 *Social Conflict: Escalation, Stalemate, and Settlement,* Second Edition. New York: McGraw-Hill.
Ruckelshaus, W.D., and K. Hausker
1998 *The Environmental Protection System in Transition: Toward a More Desirable Future.* (Final Report of the Enterprise for the Environment. Panel Report published in cooperation with the National Academy of Public Administration and The Keystone Center.) Washington, DC: Center for Strategic and International Studies.
Russo, M.V., and P.A. Fouts
1997 A resource-based perspective on corporate environmental performance and profitability. *Academy of Management Journal* 40:534-559.
Saaty, T.L.
1980 *The Analytical Hierarchy Process.* New York: McGraw Hill.
1991 *Multicriteria Decision Making: The Analytical Hierarchy Process,* Extended Edition. Pittsburgh, PA: RWS Publishers.
Sabatier, P.
1978 The acquisition and utilization of technical information by administrative agencies. *Administrative Science Quarterly* 23:396-417.
Sabatier, P., ed.
1999 *Theories of the Policy Process.* Boulder, CO: Westview Press.
Sarewitz, D., and R. Pielke, Jr.
1999 Prediction in science and policy. *Technology in Society* 21(2):121-133.
Sarewitz, D., R.A. Pielke, Jr., and R. Byerly, eds.
2000 *Prediction: Science, Decision-Making, and the Future of Nature.* Washington, DC: Island Press.
Sarokin, D.J., W.R. Muir, C.G. Miller, and S. Sperber.
1986 *Cutting Chemical Wastes: What 29 Organic Chemical Plants Are Doing to Reduce Hazardous Wastes.* New York: Inform Inc.
Schelhas, J.
2003 New trends in forest policy and management: An emerging postmodern approach? Pp. 17-27 in *Forest Policy for Private Forestry: Global and Regional Challenges,* L. Teeter, B. Cashore, and D. Zhang, eds. Cambridge, MA: CABI Publishing.

Schultz, P.W., C. Shriver, J.J. Tabanico, and A.M. Khazian
2004 Implicit connections with nature. *Journal of Environmental Psychology* 24:31-42.

Schwartz, P.
1996 *The Art of the Long View: Planning for the Future in an Uncertain World.* New York: Doubleday.

Scott, R.A., and A.R. Shore
1979 *Sociology Does Not Apply: A Study of the Use of Sociology in Public Policy.* New York: Elsevier.

Scott, W.R.
1992 *Organizations: Rational, Natural, and Open Systems.* Englewood Cliffs, NJ: Prentice-Hall.

Sexton, K., and J. Adgate
1999 Looking at environmental justice from an environmental health perspective. *Journal of Exposure Analysis and Environmental Epidemiology* 9:3-8.

Sexton, K., L. Waller, R.B. McMaster, G. Maldonado, and J. Adgate
2002 The importance of spatial scale for environmental health policy and research. *Human and Ecological Risk Assessment* 8(1):109-125.

Shapiro, C., and H.R. Varian
1999 *Information Rules.* Cambridge, MA: Harvard Business School Press.

Shapiro, S.A., and T. McGarity
1991 Not so paradoxical: The rationale for technology-based regulation. *Duke Law Journal* 40:729.

Sharma, S.
2002 What really matters: Research on corporate sustainability. Chapter 1 in *Research on Corporate Sustainability: The Evolving Theory and Practice of Organizations in the Natural Environment*, S. Sharma and M. Starik, eds. Northampton, MA: Edward Elgar Academic Publishing.

Shorey, E., and T. Eckman
2000 *Appliances and Global Climate Change: Increasing Consumer Participation in Reducing Greenhouse Gases.* Arlington, VA: Pew Center on Global Climate Change.

Shove, E., L. Lutzenhiser, S. Guy, B. Hackett, and H. Wilhite
1998 Energy and social systems. Pp. 201-234 in *Human Choice and Climate Change*, S. Rayner and E. Malone, eds. Columbus, OH: Battelle Press.

Simon, H.
1990 Invariants of human behavior. *Annual Review of Psychology* 41:1-19.

Skitka, L., and F. Crosby
2003 New and current directions in justice theorizing and research. *Personality and Social Psychology Review* (special issue) 7(4):282-399.

Slovic, P.
1995 The construction of preference. *American Psychologist* 50:364-371.

Slovic, P., and R. Gregory
1999 Risk analysis, decision analysis and the social context for risk decision making. Pp. 353-365 in *Decision Science and Technology: Reflections on the Contributions of Ward Edwards*, J. Shanteau, B.A. Mellors and D.A. Shum, eds. Boston, MA: Kluwer Academic.

Slovic, P., B. Fischhoff, and S. Lichtenstein
1977 Behavioral decision theory. *Annual Review of Psychology* 28:1-39.

Smelser, N., and R. Swedberg
1995 *Handbook of Economic Sociology.* Princeton, NJ: Princeton University Press.

Socolow, R.
1994 Six perspectives from industrial ecology. Pp. 3-18 in *Industrial Ecology and Global Change,* R. Socolow, C. Andrews, F. Berkhout, and V. Thomas, eds. Cambridge, England: Cambridge University Press.
Socolow, R., C. Andrews, F. Berkhout, and V. Thomas, eds.
1994 *Industrial Ecology and Global Change.* Cambridge, England: Cambridge University Press.
Solomon, B.D., and R. Lee
2000 Emissions trading systems and environmental justice. *Environment* 42(8):32-45.
Solomon, L.S.
1998 Evaluation of environmental programs: Limitations and innovative applications. In *Environmental Program Evaluation: A Primer,* J.G. Knaap and J.K. Tschangho, eds. Chicago: University of Illinois Press.
Stavins, R.N.
2003 *Market-Based Environmental Policies: What Can We Learn From U.S. Experience (and Related Research)?* (Regulatory Policy Program Working Paper RPP-2003-07.) Cambridge, MA: Center for Business and Government, John F. Kennedy School of Government, Harvard University.
Steel, B.S., ed.
1997 *Public Lands Management in the West.* Westport, CT: Praeger.
Stern, P.C.
2000 Toward a coherent theory of environmentally significant behavior. *Journal of Social Issues* 56:407-424.
Stern, P.C., and G.T. Gardner
1981 The place of behavior change in managing environmental problems. *Zeitschrift für Umweltpolitik* 2:213-239.
Stern, P.C., E. Aronson, J.M. Darley, D.H. Hill, E. Hirst, W. Kempton, and T.J. Wilbanks
1986 The effectiveness of incentives for residential energy conservation. *Evaluation Review* 10:147-176.
Stern, P.C., T. Dietz, T. Abel, G.A. Guagnano, and L. Kalof
1999 A value-belief-norm theory of support for social movements: The case of environmental concern. *Human Ecology Review* 6:81-97.
Stewart, R.B.
2001 A new generation of environmental regulation? *Capital University Law Review* 29(21).
Stockwell, J.R., J.W. Sorenson, J.W. Eckert, Jr., and E.M. Carreras
1993 The U.S. EPA geographic information system for mapping environmental releases of toxic chemical release inventory (TRI) chemicals. *Risk Analysis* 13:155-164.
Suchman, E.
1967 *Evaluative Research.* New York: Russell Sage Foundation.
Suter, G.
1993 *Ecological Risk Assessment.* Boca Raton, FL: Lewis Publishers, an imprint of CRC Press.
Szanton, P.
1981 *Not Well Advised.* New York: Russell Sage Foundation.
Szasz, A., and M. Meuser
1997 Environmental inequalities: Literature review and proposals for new directions in research and theory. *Current Sociology* 45(3):99-120.
Tanner, C.
1999 Constraints on environmental behavior. *Journal of Environmental Psychology* 19:145-157.

Tanner, C., F.G. Kaiser, and S.W. Kast
2004 Contextual conditions of ecological consumerism: A food-purchasing survey. *Environment and Behavior* 36:94-111.
Teuber, G.
1983 Substantive and reflexive elements in modern law. *Law and Society Review* 17(239).
Thøgerson, J.
2002 Promoting "green" consumer behavior with eco-labels. Pp. 83-104 in National Research Council, *New Tools for Environmental Protection: Education, Information, and Voluntary Measures,* Committee on the Human Dimensions of Global Change, T. Dietz and P.C. Stern, eds. Division of Behavioral and Social Sciences and Education. Washington, DC: National Academy Press.
2004 A cognitive dissonance interpretation of consistencies and inconsistencies in environmentally responsible behavior. *Journal of Environmental Psychology* 24: 93-103.
Tietenberg, T.
1998 Disclosure strategies for pollution control. *Environmental and Resource Economics* 11:587-602. Washington, DC: U.S. Congress Office of Technology Assessment.
1992 *Green Products by Design.* Washington, DC: US Government Printing Office.
2002 The tradable permits approach to protecting the commons: What have we learned? Pp. 197-232 in National Research Council, *The Drama of the Commons,* Committee on the Human Dimensions of Global Change, E. Ostrom, T. Dietz, N. Dolsak, P.C. Stern, S. Stonich, and E.U. Weber, eds. Division of Behavioral and Social Sciences and Education. Washington, DC: National Academy Press.
Travis, C.
1988 *Carcinogen Risk Assessment.* New York: Plenum Press.
Turner, J.C.
1991 *Social Influence.* Pacific Grove, CA: Brooks-Cole Publishing Company.
Tversky, A., and D. Kahneman
1981 The framing of decisions and the psychology of choice. *Science* 211:453-458.
Tyler, T.R., and H.J. Smith
1998 Social justice and social movements. Pp. 595-629 in *Handbook of Social Psychology,* Fourth Edition, Volume 2. D.T. Gilbert, S.T. Fiske, and G. Lindzey, eds. New York: McGraw-Hill.
United Church of Christ
1987 *Toxic Wastes and Race: A National Report on the Racial and Socioeconomic Characteristics of Communities with Hazardous Waste Sites.* New York: United Church of Christ, Commission for Racial Justice.
U.S. Climate Change Science Program and Subcommittee on Global Change Research
2003 *Strategic Plan for the U.S. Climate Change Science Program.* Washington, DC: U.S. Climate Change Science Program and Subcommittee on Global Change Research. Available: HtmlResAnchor http://www.climatescience.gov/Library/stratplan2003/default.htm [March 2005].
U.S. Congress Office of Technology Assessment
1992 *Green Products by Design.* Washington, DC: US Government Printing Office.
U.S. Environmental Protection Agency
1994 *Waste Prevention, Recycling, and Composting Options: Lessons from 30 Communities.* (EPA530-R-92-015.) Washington, DC: U.S. Environmental Protection Agency.
1997 *Community-Based Environmental Protection: A Resource Book for Protecting Ecosystems and Communities.* Washington, DC: U.S. Environmental Protection Agency, Office of Policy, Planning, and Evaluation.

2001 *Improved Science-Based Environmental Stakeholder Processes.* (EPA-SAB-EC-COM-01-006.) Washington, DC: U.S. Environmental Protection Agency.

2002a *Community, Culture and the Environment: A Guide to Understanding a Sense of Place.* Washington, DC: U.S. Environmental Protection Agency, Office of Water.

2002b *Innovating for Better Environmental Results: A Strategy to Guide the Next Generation of Innovation at EPA.* Washington, DC: U.S. Environmental Protection Agency.

2003 *Draft Report on the Environment 2003.* (Report No. EPA-260-R-02-006.) Washington, DC: U.S. Environmental Protection Agency.

U.S. General Accounting Office

1983 *Siting of Hazardous Waste Landfills and their Correlation with Racial and Economic Status of Surrounding Communities.* Washington DC.: U.S. Government Printing Office.

2004 *Individual Fishing Quotas: Methods for Community Protection and New Entry Require Periodic Evaluation.* (Report to Congressional Requesters, GAO-04-277.) Washington, DC: U.S. General Accounting Office.

U.S. Green Building Council

2003 *Building Momentum: National Trends and Prospects for High-Performance Green Buildings.* (Report prepared for the U.S. Senate Committee on Environment and Public Works.) Available: http://www.usgbc.org/Docs/Resources/043003_hpgb_whitepaper.pdf [March 29, 2004.]

U.S. Nuclear Regulatory Commission

1975 *Reactor Safety Study.* (USNRC, WASH 1400.) Washington, DC: U.S. Nuclear Regulatory Commission.

U.S. Nuclear Waste Technical Review Board

1995 *Report to the U.S. Congress and the Secretary of Energy: 1995 Findings and Recommendations.* Washington, DC: U.S. Nuclear Waste Technical Review Board.

U.S. Office of Management and Budget

1996 *Economic Analysis of Federal Regulations Under Executive Order 12866.* Available: http://www.whitehouse.gov/omb/inforeg/riaguide.html [December 15, 2004].

2002 *The President's Management Agenda: FY 2002.* Washington, DC: U.S. Government Printing Office. Available: http://www.whitehouse.gov/omb/budget/fy2002/mgmt.pdf [March 18, 2004].

2003 *Informing Regulatory Decisions: 2003 Report to Congress on the Costs and Benefits of Federal Regulations and Unfunded Mandates on State, Local and Tribal Entities.* Washington, DC: U.S. Office of Management and Budget.

U.S. Office of Science and Technology Policy

1983 *General Comments on Acid Rain: A Summary of the Acid Rain Peer Review Panel for OSTP.* Washington, DC: U.S. Government Printing Office.

U.S. Office of Technology Assessment

1984 *Acid Rain and Transported Air Pollutants: Implications for Public Policy.* Washington, DC: U.S. Government Printing Office.

Uusitalo, L.

1986 *Environmental Impacts of Consumption Patterns.* London, England: Gower.

Valente, T.W., and D.V. Schuster

2002 The public health perspective for communicating environmental issues. Pp. 105-124 in National Research Council, *New Tools for Environmental Protection: Education, Information, and Voluntary Measures,* Committee on the Human Dimensions of Global Change, T. Dietz and P.C. Stern, eds. Division of Behavioral and Social Sciences and Education. Washington, DC: National Academy Press.

van Asselt, M.
2000 *Perspectives on Uncertainty and Risk*. Dordrecht, The Netherlands: Kluwer.
van Heel, O.D., J. Elkington, S. Fennell, and V.D. Franceska
2001 *Buried Treasure: Uncovering the Business Case for Corporate Sustainability*. London, England: SustainAbility.
Vandenburgh, M.P.
2004 From smokestack to SUV: The individual as regulated entity in the new era of environmental law. *Vanderbilt Law Review* 57:515-628.
Vig, N.J., and M.E. Kraft, eds.
2003 *Environmental Policy: New Directions for the Twenty-First Century*. Washington, DC: CQ Press.
Vlek, C.
2000 Essential psychology for environmental policy making. *International Journal of Psychology* 35:153-167.
Vogel, D.
1995 *Trading Up: Consumer and Environmental Regulation in a Global Economy*. Cambridge, MA: Harvard University Press
2003 The hare and the tortoise revisited: The new politics of consumer and environmental regulation in Europe. *British Journal of Political Science* 33(4):557-580.
von Winterfeldt, D., and W. Edwards
1986 *Decision Analysis and Behavioral Research*. New York: Cambridge University Press.
Vringer, K., and K. Blok
1995 The direct and indirect energy requirements of households in the Netherlands. *Energy Policy* 23(10):893-910.
Wack, P.
1985a Scenarios: Uncharted waters ahead. *Harvard Business Review* 63(5):72-89.
1985b Scenarios: Shooting the rapids. *Harvard Business Review* 63(6):139-150.
Wade, R.
1994 *Village Republics: Economic Conditions for Collective Action in South India*. San Francisco, CA: ICS Press.
Walters, C.
1997 Challenges in adaptive management of riparian and coastal ecosystems. *Conservation Ecology* 1(2). Available: http://www.consecol.org/vol1/iss2/art1 [June 1, 2004].
Weber, M.L.
2002 *From Abundance to Scarcity: A History of U.S. Marine Fisheries Policy*. Washington, DC: Island Press.
Webler, T.
1995 "Right" discourse in citizen participation: An evaluative yardstick. Pp. 35-86 in *Fairness and Competence in Citizen Participation: Evaluating New Models for Environmental Discourse*. Dordrecht, The Netherlands: Kluwer.
Webler, T., S. Tuler, and R. Krueger
2002 What is a good public participation process? Five perspectives from the public. *Environmental Management* 27:435-450.
Werner, C.M., and D. Adams
2001 Changing homeowners' behaviors involving toxic household chemicals: A psychological, multilevel approach. *Analyses of Social Issues and Public Policy* 1:1-32.

Weyant, J., O. Davidson, H. Dowlatabadi, J. Grubb Edmonds, E.A. Parson, R. Richels, J. Rotmans, P.R. Shukla, R.S.J. Tol, W. Cline, and S. Fankhauser

1996 Integrated assessment of climate change: An overview and comparison of approaches and results. Pp. 367-396 in *IPCC, Climate Change 1995: Economics and Social Dimensions of Climate Change*. Cambridge, England: Cambridge University Press.

Wilbanks, T.J., and R.W. Kates

1999 Global change in local places: How scale matters. *Climatic Change* 43:601-628.

Wilk, R.

1996 *Economies and Cultures: Foundations of Economic Anthropology*. Boulder, CO: Westview Press.

Wilkinson, C.F.

1992 *Crossing the Next Meridian: Land, Water, and the Future of the West*. Washington, DC: Island Press.

Wilson, D.C., J.R. Nielsen, and P. Degnbol, eds.

2003 *The Fisheries Co-Management Experience: Accomplishments, Challenges and Prospects*. Dordrecht, The Netherlands: Kluwer.

Wilson, J.

2002 Scientific uncertainty, complex systems, and the design of common-pool institutions. Pp. 327-359 in National Research Council, *The Drama of the Commons*. Committee on the Human Dimensions of Global Change, E. Ostrom, T. Dietz, N. Dolsak, P.C. Stern, S. Stonich, and E.U. Weber, eds. Division of Behavioral and Social Sciences and Education. Washington, DC: National Academy Press.

World Resources Institute

2003 *World Resources 2002-2004: Decisions for the Earth: Balance, Voice, and Power*. Washington, DC: United Nations Development Programme, United Nations Environmental Programme and World Bank, World Resources Institute.

Yandle, T., and D. Burton

1996 Reexamining environmental justice: A statistical analysis of historical hazardous waste landfill siting patterns in metropolitan Texas. *Social Science Quarterly* 77(3):477-492.

Young, M.D., and B.J. McCay

1995 Building equity, stewardship, and resilience into market-based property rights systems. Pp. 87-102 in *Property Rights and the Environment: Social and Ecological Issues*, S. Hanna and M. Munasinghe, eds. Washington, DC: World Bank.

Young, O.R.

2002 Institutional interplay: The environmental consequences of cross-scale interactions. Pp. 263-292 in National Research Council, *The Drama of the Commons*, Committee on the Human Dimensions of Global Change, E. Ostrom, T. Dietz, N. Dolsak, P.C. Stern, S. Stonich, and E.U. Weber, eds. Division of Behavioral and Social Sciences and Education. Washington, DC: National Academy Press.

Zadeh, L.

1965 Fuzzy sets. *Information and Control* 8:338-353.

Zerbe, R.O., Jr. and D.D. Dively

1994 *Benefit-Cost Analysis in Theory and Practice*. New York: Harper-Collins.

Zimmerman, R.

1994 Issues of classification in environmental equity: How we manage is how we measure. *Fordham Law Journal* 21:633-670.

Appendix A

When Do Environmental Decision Makers Use Social Science?

Rebecca J. Romsdahl

This appendix describes a body of literature that is relevant to understanding the conditions under which decision makers are likely to use social science information in environmentally significant decisions. The literature was identified by contacting selected researchers and searching numerous databases (AGRICOLA, BIOSIS Previews, CSA Environmental Science and Pollution Management, EconLit, Elsevier-ScienceDirect, EMBASE, InfoTrac OneFile, National Technical Information Service, ProQuest General Reference, HtmlResAnchor PsycINFO, Public Affairs Information Service [PAIS International], Social Sciences Citation Index, Sociological Abstracts) using combinations of the following topics: environmental, decision making, policy making, social science, knowledge utilization, and information utilization. The search was not exhaustive, but it is presented as representative of the field.[1] This appendix characterizes the literature, lists some of the reoccurring conclusions found in the studies, and concludes with an annotated bibliography containing 54 citations. The annotations summarize the following questions for each article:

- What is the empirical basis, if any?
- What is the social science and environmental domain?
- Who are the decision makers involved?
- What are the conclusions, if any, on how social science was/is/would be used?
- Are there recommendations or any other relevant pieces of information?

Very few of the studies directly address the use of social science information in environmental decision making. The majority are from fields of social policy (e.g., education, health), but a few provide examples from other fields. One study (Rosen, 1977) reviews literature on the use of social science in judicial policy making; this study provides interesting but limited insight in this area of decision making. Another limited study (Deshpande, 1981) addresses the use of social science research in business decisions. Given the focus of this bibliography on the utilization of social science, no attempt was made to summarize studies on the use of natural science in decision making; however, one illustrative study is included (Powell, 1999). This study addresses the use of natural science information, in the U.S. Environmental Protection Agency, that is of direct relevance to regulatory decision making. In addition, although it is recognized that claims are often made about the misuse of scientific information in government decision making, the time frame of this review precluded a search for literature to examine such claims in regard to social science information. Most of the literature in this bibliography comes from the study of "knowledge utilization," a popular research area in the 1970s and early 1980s. Interest in this field of research seems to rise and fall periodically and it has been less active in recent years; however, as this panel study shows, questions about social science utilization persist and recent studies do build on and advance earlier work.

SOCIAL SCIENCE UTILIZATION IN GOVERNMENT DECISION MAKING

The most recent studies in this field (Landry, Amara, and Lamari, 2001; Landry, Lamari, and Amara, 2003) are significant for their broad examination of decision-making offices, including social and environmental, and their critique of the knowledge utilization literature. In their analyses, Landry et al. (2001, 2003) consider organizational and communication factors and find that both influence utilization. For example, they highlight the importance of policy domain: university research reached its highest levels of utilization in the fields of education and information technology (Landry et al., 2003).

Many of the early studies on knowledge utilization focus on federal government decision makers primarily in areas of social policy. The studies reviewed here present some useful insights into how government decision makers use social science information. Among these are practical typologies of social science roles in decision-making processes. It is useful for the present purpose to highlight two broad categories:

- the conceptual or enlightenment role—social science providing a broad information base for decisions (Caplan, Morrison, and Stambaugh, 1975; Dunn, 1983; Nelson, Roberts, Maederer, Wertheimer, and Johnson, 1987; Oh, 1996a; Oh and Rich, 1996; Patton et al., 1977; Pollard, 1987; Weiss, 1977, 1979; Weiss and Bucuvalas, 1977, 1980; Wilensky, 1997)
- the instrumental role—information put to use for specific decisions or requested by decision makers for specific projects (Deshpande, 1981; Knorr, 1977; Oh, 1996b; Weiss, 1979)

The literature also identifies other roles, including justifying or legitimating decisions already reached (Caplan et al., 1975; Knorr, 1977; Oh, 1996b; Scott and Shore, 1979; Weiss, 1979) and serving as a substitute for or justification for postponing actual decisions (Knorr, 1977; Oh, 1996b; Scott and Shore, 1979; Weiss, 1979).

Several studies address how information comes to be utilized in these roles by exploring two major competing hypotheses (Greenberg and Mandell, 1991; Majchrzak, 1986; Oh, 1996a, 1996b, Oh and Rich, 1996; Rich, 2001). One hypothesis focuses on the characteristics of the information: if the information is "relevant, timely, and comprehensible, it will be used" (Majchrzak, 1986). The other focuses on organizational or bureaucratic factors, suggesting, for instance, that information will be used "when the rewards and incentives of the organizational structure encourage its use" (Majchrzak, 1986) or when the information is consistent with the ideology and interests of the organization and/or its members (Weiss, 1983).

Other studies identify additional characteristics of the most frequently utilized information:

- it is in the form of social statistics (Caplan, 1976)
- it comes from internal agency sources (Caplan et al., 1975; Nelson et al., 1987; Oh, 1996a, 1997; Oh and Rich, 1996)
- it supports decisions that have been made on other grounds (Knorr, 1977; Oh, 1996b; Scott and Shore, 1979)
- it is perceived to support the decision-maker's perception of the agency's best interests (Oh, 1996a)
- it provides means to improve the sponsoring agencies' bureaucratic efficiency (Caplan, 1976; Scott and Shore, 1979)
- it was specifically requested by the decision maker (Caplan, 1976; Landry et al., 2001, 2003)

In addition to reviewing these broadly based studies, the search extended to documents from U.S. government agencies responsible for natural resource management. The bibliography includes two studies that examine social science utilization by such agencies—the National Oceanic and Atmospheric Administration (NOAA) and the Minerals Management Service (MMS) under the U.S. Department of the Interior.

NOAA recently published a report on the findings of its external social science review panel. Two general findings were presented: (1) "The capacity of NOAA to meet its mandates and mission is diminished by the under-representation and under-utilization of social science" and (2) "Assistant Administrators are responsive to discussing opportunities for an enhanced role for social science within their line offices" (NOAA, 2003). The report also compares NOAA to several other regulatory environmental agencies and finds it lacking. "The line office budgets for social science research, education and staffing do not seem comparable to the social science budgets at other agencies with environmental assessment and stewardship responsibilities such as U.S. Forest Service, the Environmental Protection Agency, or the U.S. Fish and Wildlife Service" (NOAA, 2003). Overall, the report concludes that "the position of social science within NOAA is weak" but it presents an array of recommendations for improving the use of social science research in the agency (NOAA, 2003). Some of these recommendations include having headquarters and each line office develop social science research plans that identify goals and implementation strategies to help the agency accomplish its mission, creating a chief social scientist position in each line office, and using external experts to help educate personnel about potential contributions of social science to NOAA's goals.

The MMS report (Luton and Cluck, 2000) is an internal assessment conducted by two social scientists employed by the service. The authors find that

> the MMS uses social science data and analysis throughout the various phases of decisionmaking: 5-year planning, prelease and leasing activities, exploration, development, production, and decommissioning of offshore platforms. The MMS designs studies to address the data and analytical needs arising from these specific phases in order to aid in the decision-making process.

The researchers also describe eight broad categories of social and economic research components that are used by the service, including issues identification, national economic analysis, and community- and individual-level analysis. For each category they identify the data needs and level of detail required in order for research in that category to support decisions at the various stages of policy making.

Other federal natural resource agencies were contacted including the U.S. Forest Service, which employs social scientists but appears to be examining its use of science in a much broader sense at present; the National Park Service, which employs a visiting chief social scientist and has an ongoing national program in social science research, but application of that research to decision making is described as decentralized[2]; the Bureau of Land Management, which employs social scientists and has a chief social scientist but has not conducted a broad assessment of its social science utilization; and the Fish and Wildlife Service, which does not appear to have a chief social scientist and has not responded to inquiries.

CHALLENGES TO SOCIAL SCIENCE UTILIZATION

Some of the studies present barriers to utilization that might be overcome by actions social scientists can take:

- failure to produce results in the form of generalized principles or politically feasible recommendations (Boggs, 1990; Caplan et al., 1975; Freudenburg, 1989; Freudenburg and Keating, 1985; Greenberg and Mandell, 1991; Jones, Fischhoff, and Lach, 1999; Scott and Shore, 1979; Useem, 1977; Weiss and Bucuvalas, 1980)
- lack of clarity on research questions and/or policy objectives (Corwin and Louis, 1982; Freudenburg, 1989; Fricke, 1985; Jones et al., 1999; Rich, 2001)
- disagreement on findings, i.e., a lack of consistent or cumulative research results on a given subject (Gismondi, 1997; Lindblom and Cohen, 1979; Weiss and Bucuvalas, 1980)

Some findings identify barriers that social science is unlikely to address; some of these might be addressed by changes in the organizations that use social science information:

- a lack of clear roles for scientists in decision-making processes (Boggs, 1990; Freudenburg, 1989; Webber, 1987)
- political influences (Corwin and Louis, 1982; Freudenburg, 1989; Freudenburg and Keating, 1985; Gismondi, 1997; Patton et al., 1977)
- unavailability of social science research results until after a decision must be made (Dreyfus, 1977; Greenberg and Mandell, 1991; Healy and Ascher, 1995; Jones et al., 1999; Weiss and Bucuvalas, 1980)
- low credibility of social science information relative to natural science information (Gismondi, 1997; Sabatier, 1978)

Some other research conclusions are also worth mentioning. Weiss (1977) found that decision makers in the mental health field were open to controversial research that made them reassess comfortable assumptions; these decision makers found it possible that others in their fields would also consider such research in their decision-making processes. Weiss and

Bucuvalas (1980) found four situations when decision makers in the mental health field commonly sought new information:

- when they faced new circumstances
- when they had to make decisions that involved important or expensive outcomes
- when they might request consultants' help on issues where they lacked expertise
- in situations where they wanted authoritative backup because their judgment might be challenged

In a study of legislative decision makers, Webber (1987) found that most were unlikely to use policy information or social science if left to their own tendencies. Legislators were more likely to use these sources if they already viewed social science as valid and useful information; if the research supported views they already held; or if their constituents had requested such information, asked questions about it, or demanded that attention be paid to issues covered by it.

STUDIES IN ENVIRONMENTAL DECISION MAKING

In environmental decision making, there is a great deal more research on the use of natural science than social science. A quick search through the database Elsevier-ScienceDirect, on science and environmental policy, for example, will bring up dozens of articles. These range in topic from examinations of risk assessment and scientific uncertainty in policy making to incorporating long-term monitoring and environmental impact assessments. No attempt was made to review this literature.

Studies of social science use in environmental decision making are sparse overall and tend to focus on case studies of social impact assessment, but a couple of insights are worth mentioning. Freudenburg (1989) highlights that social scientists must overcome the hurdle of explaining to nonsocial science background persons the many ways in which environmental policies are social and the need for environmental decision making to use social science information. Fricke (1985) discusses an important factor in environmental decisions that is also mentioned in articles on social policy—the need for better communication between researchers and decision makers before research begins in order to clarify the objectives of and required knowledge for projects and planning.

INSIGHTS FROM SOCIAL STUDIES OF SCIENCE

Although this field of research would likely add another valuable perspective to the analysis of social science utilization, this review examined it only briefly, as it surfaced late in the study time frame. One example where this field of study could provide valuable insights is in understanding how researchers interact with those who will be using their research findings. Freudenberg and Gramling (2002) discuss the variety of ways that natural scientists often struggle to remain unbiased in conducting their work, especially when they are asked to provide information in policy-making situations.

The authors provide an insightful analysis of Paul Hirt's 1994 book, *Conspiracy of Optimism*. In examining the U.S. Forest Service policy for promoting "sustained yield" of wood production, the authors' focus on Hirt's conclusion that even when Forest Service scientists were committed to carrying out balanced research and believed that they were doing so, their findings often resulted in significant short-term benefits for those interests that were focused on exploitation of the resource over the broader interests of the resource and the public interest. Freudenberg and Gramling (2002) explain this phenomenon in terms of how the research process can be limited by blind spots and scientific limitations. "Few of those scientists have had any difficulty in recognizing this pattern in retrospect; equally few of them, unfortunately, appear to have been able to recognize it in advance. The authors go on to suggest that natural science researchers in this situation may have benefited from interaction with social scientists, especially those who would be familiar with "unseen, structural biasing pressures" present in many research scenarios. Social science analyses of the relationships between researchers and those requesting the research would likely benefit not only from study outcomes but also their utilization.

RECOMMENDATIONS FROM THE REVIEWED LITERATURE

What Social Science Researchers Can Do to Improve Utilization

Some of the studies make recommendations for improving utilization of social science by government decision makers. Francis, King, and Riddlesperger (1980) suggest that researchers should "[target] evaluations to the interests of the administrators or legislators, [and] use an appropriate justification when suggesting programmatic or policy changes." In contrast, Landry et al. (2001, 2003) find that focusing research on users' needs does not improve utilization any more than research focused on the advancement of scholarly knowledge. Other studies suggest that

- researchers should explore alternative approaches and roles in policy making, such as forming groups who can translate university research into policy recommendations or translate policy issues into researchable questions, seeking appointment to science policy committees or encouraging interest groups to push for committee members who will listen to social scientists (Boggs, 1990; Caplan, 1977; Catalano, Simmons, and Stokols, 1975; Freudenburg and Gramling, 2002; Freudenburg, 1989)

- social scientists' policy recommendations should be based on appropriate political factors (Caplan, 1977; Patton et al., 1977)

Dissemination of social science research is presented as an important positive influence on utilization in at least four studies. One study (Huberman, 1990) finds that greater contacts, including face-to-face and follow-up interactions, between researchers and decision makers throughout a study often lead to increased promotion and distribution of research findings in later stages of the dissemination process. Another study (Greenberg and Mandell, 1991) states that results might be underutilized if researchers do not take the initiative to distribute their studies directly to practitioners.

The most recent studies (Landry et al., 2001, 2003) also suggest that utilization can be increased by emphasizing links between researchers and decision makers and by encouraging researchers to take the initiative in dissemination so that research is more widely available to decision makers. One example is to compensate or reward researchers for the costs of directly distributing their research. Some researchers recommend that government decision makers take some responsibility in this process by actively involving social scientists at the beginning of planning projects (Fricke, 1985; Gans, 1971; Gismondi, 1997).

CONCLUSIONS

This appendix presents a brief summary of the state of knowledge in the field of social science utilization. The references included in the annotated bibliography were chosen for their broad representation of this field and their applicability to the question of how decision makers use social science. For additional references, see Landry et al. (2003); this article appears to be the most current analysis of the knowledge utilization literature and its application to government decision making.

The literature suggests actions that can be taken by both sides to expand the use of social science research. Researchers can take the initiative to meet with policy makers at regular intervals during the research process or directly distribute their findings to those policy makers who might be able to utilize the research. Policy makers and research funders can provide incentives to encourage social science researchers to be more proactive in distributing their research and to consider political factors when making policy recommendations. Policy makers can also take the initiative to more actively involve social scientists at the beginning stages of planning projects.

The field of social science utilization is one that would greatly benefit from additional research. "We know little about the factors that induce professionals and managers in government agencies to use university research in their professional activities" (Landry et al., 2003). The question of how social science is used in government decision making should not simply be an academic pursuit; government agencies' use or nonuse of social science information has significant impacts on the lives of citizens as officials make decisions and create policies. This is only one reason why social science research on science utilization can contribute to better environmental decision making.

NOTES

1. For instance, the search did not include variants of the phrase "evidence-based decision making."

2. For additional information see the National Park Service web site: HtmlResAnchor http://www.nature.nps.gov/socialscience/index.htm (last visited June 2004).

ANNOTATED BIBLIOGRAPHY

Boggs, J.P.

1990 The use of anthropological knowledge under NEPA. *Human Organization* 49(3):217-226.

Notes:

- Literature review and analysis
- Use and influence of social science under the National Environmental Policy Act (NEPA) was examined
- Decision makers were at the federal level
- Conclusions: The study found that social science fails to produce results in the form of generalized principles that can be applied to particular cases. Practitioners need closer, more effective open links with basic social sciences.
- Recommendation: The author suggested the development of a professional role for social science under NEPA that is grounded in the basic social sciences.

Caplan, N.

1976 Social research and national policy: What gets used, by whom, for what purposes, and with what effects? *International Social Science Journal* 28(1):187-194.

Notes:

- 204 face-to-face, recorded interviews (dataset from Caplan et al., 1975, report)

• Social science research utilization and policy formation were examined

• Decision makers represented high-level civil servants or political appointees from across the entire range of government activities

• Conclusions: The study found that social statistics were the most frequently used data. Most of the information used in a decision was sponsored by the deciding agency, and most of the knowledge utilization was applied toward improving bureaucratic efficiency. Policy makers' information processing style (clinical, academic, or advocacy orientation) stood out as having special influence on the level of their utilization. Policy makers also emerged as playing an active role in prescribing the information that they wanted and would ultimately use.

1977 A minimal set of conditions necessary for the utilization of social science knowledge in policy formulation at the national level. In *Using Social Research in Public Policy Making*, C.H. Weiss, ed. (Policy Studies Organization Series.) Lexington, MA: Lexington Books.

Notes:

• 204 face-to-face, recorded interviews (dataset from Caplan et al., 1975, report)

• Social science research utilization and policy formation were examined

• Decision makers represented high-level civil servants or political appointees from across the entire range of government activities

• Conclusions: The study found that the most frequent reason given for nonutilization of relevant social science information was that the implications are politically unfeasible. So, to increase utilization, the gap between social scientists and policy makers' perspectives must be bridged, but there does not necessarily need to be more direct contacts.

• Recommendations: The author suggested the formation of a group of individuals representing different roles and skills in research and policy making who can make realistic and rational appraisals of available social science information, make appropriate translations from university research to policy-making situations, recast policy issues into researchable terms, identify and distinguish between scientific and "extrascientific" knowledge needs, deal with the value issues and bureaucratic factors that influence both the development and the use of scientific results, and gain policy makers trust and sufficient understanding of the policy process in order to introduce social science in ways that will increase its utilization.

Caplan, N., A. Morrison, and R.J. Stambaugh

1975 *The Use of Social Science Knowledge in Policy Decisions at the National Level: A Report to Respondents.* Ann Arbor: University of Michigan.

Notes:

- Statistical analysis of 204 face-to-face recorded interviews
- Social science research utilization and policy formation were examined
- Decision makers represented high-level civil servants or political appointees from across the entire range of federal government activities including environmental and natural resource management
- Conclusions: The study found that decision makers used social science in diverse ways including such examples as a basis for planning, evaluating, and determining feasibility of programs and increasing bureaucratic efficiency. Most of the information used was from internal sources or directly funded by the agency. Sociology ranked highest in frequency of use as did the methodology of program evaluation. Newspapers and government reports were the most frequently mentioned sources of social science information, with staff assistance and books listed second and professional journals third. Decision makers perceived social science information as most important in sensitizing policy makers to social needs. Factors that influenced utilization included decision makers interest and receptivity to social science information, a lack of understanding and/or mistrust between policy makers and researchers (two communities theory), a perceived objectivity of the data, findings that were counterintuitive to policy makers personal beliefs were often rejected, political feasibility, policy maker's information processing style, and policy maker's career plans—if they were unsatisfied with their position and planning to change careers, they were less likely to use social science information.

Catalano, R., S.J. Simmons, and D. Stokols

1975 Adding social science knowledge to environmental decision making. *Natural Resources Lawyer* 8(1):41-58.

Notes:

- Case study
- Environmental Impact Report process in California was examined
- Decision makers were at the state level
- Conclusions: The study suggested three ways for social scien-

tists to contribute to the environmental impact assessment process: (1) through direct participation as a citizen, (2) by teaching social science methods to professionals in environmental management, and (3) by turning research attention to the development of predictive models of social impacts for future types of environmental impact assessment projects.

Corwin, R.G., and K. Seashore Louis

1982 Organizational barriers to the utilization of research. *Administrative Science Quarterly* 27:632-640.

Notes:

- Secondary analysis of case studies, retrospective interviews
- Education demonstration programs were examined
- Decision makers were at the federal level
- Conclusions: The study found that many organizational characteristics were barriers to research utilization, including a lack of clear research questions, conflicts over and vagueness in research designs—especially tensions between theory- versus policy-driven designs, a lack of consistent policy options and objectives, high rates of personnel turnover and changing policy contexts, overlapping bureaucratic jurisdictions and interagency rivalries resulted in poor interagency coordination and cooperation, and decentralized decision making and lack of ties between policy research and long-term operational programs isolated the potential influence of information.

Deshpande, R.

1981 Action and enlightenment functions of research. *Knowledge: Creation, Diffusion, Utilization* 2(3):317-330.

Notes:

- Literature review and analysis (over 60 articles), personal interviews, and mail survey (92 respondents, all from businesses)
- Information utilization in private organizations was examined, and public and private decision making were compared
- Decision makers were policy-making private executives in consumer products and services manufacturing and distribution
- Conclusions: The study found that private organizations used instrumental information and in most cases contracted research agencies for the exact purpose of obtaining information on specific questions whereas public organizations used conceptual information and it had a more indirect influence on decision making.

Dreyfus, D.A.

1977 The limitations of policy research in congressional decision making. In *Using Social Research in Public Policy Making,* C.H. Weiss, ed. (Policy Studies Organization Series.) Lexington, MA: Lexington Books.

Notes:
- Literature review and analysis
- Policy information was examined
- Decision makers were federal-level policy makers in Congress
- Conclusions: The study found that by the time an issue reaches Congress for a decision, decision makers do not have time to consider new information on the subject. They play a summary role where the use of more information would be excessive.

Dunn, W.N.
1983 Measuring knowledge use. *Knowledge: Creation, Diffusion, Utilization* 5(1):120-133.
Notes:
- Literature review and analysis
- An inventory of concepts, procedures, and measures from social science studies of knowledge was conducted
- Decision makers were social scientists in their capacity as researchers
- Conclusions: The study found that little is known about the use of science and experiential knowledge by individuals and collectives. The author recommended areas for further research.
- Recommendations: The author suggested that the following are needed: a better understanding of the convergent and discriminant validity of constructs for the assessment of subjective properties, concepts that capture the sociocognitive complexity of knowledge use processes, examination of both the benefits of using science and professional knowledge and the drawbacks and the various reference frames and social systems of which researchers and policy makers are also members, and identification and development of concepts to distinguish the range of expected general and specific effects of knowledge use including general organizational and government learning and public enlightenment.

Florio, E., and J.R. Demartini
1993 The use of information by policy-makers at the local-community level. *Knowledge: Creation, Diffusion, Utilization* 15(1):106-123.
Notes: Full text unavailable for review
Abstract: The goal of this study was to examine how policy makers at the local community level use social science information in making decisions. The assumption that guided the study is that the use of social science information is related to how it interacts with other information and with the ideology and interests of the policy maker in the decision-making process. Findings from the study revealed that policy makers drew on a variety of information

sources. The use of social science information was dependent on the ideology and interests of the decision makers and on the specific circumstances that shaped the decision-making process.

Francis, W.L., J.D. King, and J.W. Riddlesperger
1980 Problems in the communication of evaluation research to policy makers. *Policy Studies Journal* 8:1184-1194.
Notes: Full text unavailable for review
Abstract: There are three probable causes for lack of receptivity to evaluation research: (1) a nonacceptance of scientific orientation to public policy problems, (2) critical differences between evaluators and users as to what constitutes a problem, and (3) preferences for alternative policy justifications. These causes are examined with data from an interview survey of 15 agency administrators and 15 legislators. The first cause was found to be unlikely and the last two probable. The lesson for evaluators is to key evaluations to the interests of the administrators or legislators, and to use an appropriate justification when suggesting programmatic or policy changes.

Freudenburg, W.R.
1989 Social scientists' contributions to environmental management. *Journal of Social Issues* 45(1):133-152.
Notes:
- Literature review and analysis
- Factors limiting the use of social science in environmental decisions were examined
- Decision makers were at the federal agency level
- Conclusions: The study found that the lack of social science inclusion in environmental decision making can be attributed to many factors, including social scientists limiting their own effectiveness by not communicating effectively with decision makers or failing to offer realistic suggestions for policy changes; social scientists must also overcome the hurdle of explaining to those with a nonsocial science background the many ways in which environmental policies are social, the imbalance between resources and expectations for social science results versus other sciences, and political influences may be a large factor because scientists have a limited role in policy-making processes.
- Recommendations: The author suggests that scientists should explore alternative approaches and roles in policy making and pay greater attention to political factors, especially the balance of access to scientific resources.

Freudenburg, W.R., and R. Gramling
2002 Scientific expertise and natural resource decisions: Social science participation on interdisciplinary scientific committees. *Social Science Quarterly* 83(1):120-136.
Notes:
- Literature review and analysis
- Factors limiting the use of social science in environmental decisions were examined
- Decision makers were at the federal agency level
- Conclusions: The study found that there is a need for more social science knowledge across disciplines and specifically in natural resource policy-making arenas. The authors argue that additional social science involvement could help biophysical scientists reflect on their role in the policy process and in doing so help them recognize subtle pressures that may result in biased research.
- Recommendations: The authors argue that social scientists should seek out greater roles in policy making, specifically membership on interdisciplinary scientific committees in natural resources policy making.

Freudenburg, W.R., and K.M. Keating
1985 Applying sociology to policy: Social sciences and the environmental impact statement. *Rural Sociology* 50(4):578-605.
Notes:
- Literature review and analysis
- Factors limiting the use of social science (specifically social impact assessments) in environmental impact statements were examined
- Decision makers were federal agencies
- Conclusions: The study found that the limiting factors included the overall difficulty of conducting social impact assessments, their anticipatory nature as compared to empirical data, limited funding, inertia in the discipline of social science, organizational resistance, and the political nature of the process.
- Recommendations: The authors suggested changes that could overcome the limiting factors, including cooperating with environmental or public interest groups that are litigating against the environmental impact statement; assisting state and local groups in adversarial actions; and working simultaneously for both sides in a dispute.

Fricke, P.H.
1985 The use of sociological information in the allocation of natural resources by federal agencies: A comparison of practices. *Rural Sociologist* 5(2):96-103.

Notes:

- Literature review and analysis
- Use of social impact assessments (SIAs) in natural resource decision making was examined
- No distinct decision makers were identified; instead National Forest Service and National Marine Fisheries Service (NMFS) policy processes were compared
- Conclusions: The study found that the National Forestry Service has successfully ensured the inclusion of SIAs in its policy process through its long-standing practice of developing agency-wide directives for new procedures. The NMFS has not had as much time to develop its operating procedures and was still required to work with regional management councils. These two groups could not agree on the validity of SIAs and therefore the assessments were not being incorporated into NMFS's planning processes.
- Recommendations: The author suggested that the key to successful management of common pool resources is prior agreement on objectives and knowledge required for planning and integration of all elements (social, economic, biological, etc.) at the lowest levels (i.e., the plan-development teams).

Gans, H.J.

1971 Social science for social policy. In *The Use and Abuse of Social Science,* First Edition. I.L. Horowitz, ed. New Brunswick, NJ: Transaction Books.

Notes:

- Literature review and analysis
- Social science for policy making was considered
- Decision makers were at the federal level
- Conclusions: The study found that policy making needs social science in its design stages to assist in decisions about which program will achieve the desired goals. Policy makers need as much empirical evidence as they can get to support their decisions and determine if it is possible to achieve the desired goals. Social science should provide policy makers with empirical models of all the components, stages, and consequences of alternatives. The author identified factors that make social science research problematic for policy making, including a detached researcher perspective, impersonal universalism, high generality, conceptual abstractions, metaphysical assumptions, and inattention to theories and concepts of power.

Gismondi, M.

1997 Sociology and environmental impact assessment. *Canadian Journal of Sociology-Cahiers Canadiens de Sociologie* 22(4):457-479.

Notes:

- Case study
- Environmental impact assessment (EIA) process in Canada was examined
- Decision makers were at the federal level and interested public citizens were included
- Conclusions: The study examined EIA literature and a specific EIA case. It found the following challenges to the utilization of social research: Some social questions were screened out of the EIA process by political influences, research priority setting was not open to the public early in the EIA process, social science still needs to be elevated to an equal regard with natural science, and experts often disagree on research findings about the same issue.
- Recommendations: The author suggested that social science could contribute to EIAs by quantifying the number of peer-reviewed and non-peer-reviewed research studies supporting a given proposal and assessing them for bias, context, and alternatives presented. Other recommendations included that social science could identify the extent to which natural and physical scientists bias their findings with personal value-based inputs and it could provide understanding of the social interaction of public speaking in contexts of unequal power.

Greenberg, D.H., and M.B. Mandell

1991 Research utilization in policymaking: A tale of two series (of social experiments). *Journal of Policy Analysis and Management* 10(4): 633-656.

Notes:

- Case study
- Social experiments in general and two specific projects were examined—income maintenance and welfare demonstration
- Decision makers were at the federal level
- Conclusions: The study found that one project resulted in symbolic and persuasive use—to give support to those arguing one side of the issue, while the other project resulted in more concrete elaborative use by providing input toward the development of new legislation. It also summarized, from the literature, that utilization was influenced by two sets of factors. One focused on five characteristics of the information: credibility, timeliness, communicability, and visibility, generalizability, and relevance. The other focused on characteristics of the policy environment. The study also

highlighted the importance of dissemination efforts as emphasized by the utilization literature.

Healy, R.G., and W. Ascher

1995 Knowledge in the policy process: Incorporating new environmental information in natural resources policymaking. *Policy Sciences* 28(1):1-19.

Notes:

- Literature review and analysis
- Implications of using new information (i.e., ecosystem management, valuation of ecosystem functions) in natural resource policy making were examined
- Decision makers were the National Forest Service and other government and nongovernment natural resource policy makers
- Conclusions: The study found that advances in new knowledge were expected to improve policy legitimacy, acceptance, and implementation. But new information sometimes left nonexperts more powerless to influence decisions, polarized debates over the appropriate use of resources, and delayed decisions.

Huberman, M.

1990 Linkage between researchers and practitioners: A qualitative study. *American Educational Research Journal* 27(2):363-391.

Notes:

- Multicase tracer study
- Education research and practice were examined
- Decision makers were education practitioners
- Conclusions: The study found that the greater the formal and informal contacts between researchers and practitioners during a study, the greater the collaboration afterward. In addition, the informal contacts developed during the study energized intermediaries who aggressively disseminated and promoted the use of study results.

Jones, S.A., B. Fischhoff, and D. Lach

1999 Evaluating the science-policy interface for climate change research. *Climatic Change* 43(3):581-599.

Notes:

- Case study, interviews
- Previous topic literature was examined and interviews were conducted with 14 policy makers in the Pacific Northwest salmon issue
- Decision makers were at the federal and state level
- Conclusions: The study recommended four conditions necessary for science research to be utilized in decision making: (1) Research results must be relevant to currently pending decisions.

(2) Research results must be compatible with existing policy-making processes and models. (3) Research results must be accessible to the appropriate policy makers. (4) Policy makers must be receptive to the research results.

• Recommendations: The authors recommend the use of integrated assessments to improve the utilization of science in policy making and suggest that this type of model could be adapted to other research areas as well.

Knorr, K.D.

1977 Policy makers' use of social science knowledge: Symbolic or instrumental? In *Using Social Research in Public Policy Making*, C.H. Weiss, ed. (Policy Studies Organization Series.) Lexington, MA: Lexington Books.

Notes:

• 70 face-to-face interviews

• Government-contracted social science projects were considered

• Decision makers were medium-level policy makers in Austrian federal and municipal government

• Conclusions: The study categorized four functions of social science research utilized by government: census, motivation, acquisition, and rationalization functions. It also identified four roles social science played in decision making: (1) as an information base for actual decisions, (2) as a direct translation of results into practical measures and action strategies, (3) as a substitution for an actual decision or other action, and (4) to legitimize a decision made for different reasons (this one is less common than perceived). Overall, the study found that use of social science information is characterized as diffuse, indirect, difficult to pinpoint who uses it where in the process, and as having a "delayed discursive processing" of results.

Landry, R., N. Amara, and M. Lamari

2001 Utilization of social science research knowledge in Canada. *Research Policy* 30(2):333-349.

Notes:

• Survey composed of 1,229 interviews, multiple regression analysis

• Decision makers were Canadian social science scholars

• Conclusions: The study found that the assumption of underutilization of social science by decision makers might be explained by the narrow definition of knowledge utilization limited to instrumental use. It also suggested that some researchers and decision makers have overlooked more recent empirical studies of knowledge utilization. Study found that important determinants of utilization included the mechanisms linking the researchers to the users, the dissemination efforts, the adaptation of research outputs undertaken by the researchers, and the users' context and the publication assets of the researchers.

Landry, R., M. Lamari, and N. Amara

2003 The extent and determinants of the utilization of university research in government agencies. *Public Administration Review* 63(2):192-205.

Notes:

- Survey, multiple regression
- Decision makers were 833 Canadian government officials in a broad number of agencies including environmental fields
- Conclusions: The study found that utilization cannot be explained by research characteristics, a focus on the advancement of scholarly knowledge, or on users' needs. Good predictors of research utilization included users' adaptation of research, users' acquisition efforts, links between researchers and users, and users' organizational contexts.

Lester, J.P.

1993 The utilization of policy analysis by state agency officials. *Knowledge: Creation, Diffusion,Utilization* 14(3):267-290.

Notes: Full text unavailable for review

Abstract: Findings from a 1988 survey of U.S. state officials working in the areas of hazardous wastes, economic development, welfare, and education suggest that these officials do not appear to rely heavily on policy analysis from research organizations or from university faculty; instead, they rely principally on policy advice from their peers in other state agencies, newspapers, their counterparts in federal agencies, and staff from the governors' office. In attempting to understand knowledge utilization, the study found that, among the variables considered, utilization of policy formation is best explained by state contextual variables and user characteristics That is, agency officials in wealthier, more conservative, moralistic states used policy analysis in their work more than officials in poorer, more traditional, liberal states. In addition, more experienced and better educated officials used policy advice less than inexperienced and less educated officials.

Lindblom, C.E., and D.K. Cohen

1979 *Usable Knowledge: Social Science and Social Problem Solving.* New Haven, CT: Yale University Press.

Notes:

- Literature review and analysis

• No distinct decision makers are identified, instead, the book provides a broad review of social science and problem solving in order to consider uses in government, business, and other situations.

• Conclusions: The study found that social science and research are not well understood by their own researchers and that this misunderstanding is a significant factor in the lack of utilization. The primary problem is discussed as neglect on the part of social scientists to consider the wide range of possible inputs when they study the role of social science in problem solving. Specifically, the authors stated the importance of including interactive problem solving, social learning, and ordinary knowledge in these considerations. Other problems included misplaced beliefs in authoritativeness, the high costs of social science research, the lack of conclusive answers in light of limited human cognition and complexity of the social world, wasted resources on overstudied topics, and impossible tasks assigned to overextended agencies. These have led to a situation where the authors see social problem solving as being removed from "rational problem solving."

• Recommendation: The authors suggested that social problem solving must be coupled with interactive problem solving and analysis. Social science research could be improved by combining it with social interaction.

Luton, H., and R.E. Cluck

2000 *Applied Social Science for MMS: A Framework for Decision Making*. Washington, DC: Minerals Management Service Environmental Studies Program.

Notes:

• Literature review and analysis

• Social science research in the Minerals Management Service (MMS) was examined

• Decision makers were at the federal government level

• Conclusions: The study outlined why MMS conducts social science research and described eight broad categories of ongoing social and economic research at MMS. The categories are issues identification, national economic analysis, regional-level analysis, community- and individual-level analysis, resource use issues, adaptive policy studies, mitigation, and monitoring. The study then identified the data needs and level of details required for research in these areas to support decisions in various policy stages. The study concluded that the MMS has given increased emphasis to socioeconomic research in recent years.

Majchrzak, A.

1986 Information focus and data sources: When will they lead to use? *Evaluation Review* 10(2):193-215.

Notes:

- Interviews with 90 respondents from capital cities of seven states, statistical analysis of relationships between decision types
- Social service domain was examined
- Decision makers were at the state government level
- Conclusions: The study categorized four types of decisions: performance appraisal, resource requirement, program change, and those establishing criteria for assessing effectiveness. It identified nine types of information used by the decision makers and grouped those into four categories: inputs, processes, outputs, and impacts. The study also identified eight sources of information used by the decision makers: agency archival, evaluation or special studies, reviews of client records or observations, performance reports, comparison reports, advocacy or public comments, needs assessments, and other more specific sources such as service providers. The study found that utilization was influenced by the decision maker's role in the organization and their role was related to the type of decision but not to the information focus.

Mooney, C.Z.

1992 Putting it on paper—The content of written information used in state lawmaking. *American Politics Quarterly* 20(3):345-365.

Notes: Full text unavailable for review

Abstract: What kinds of information do state legislators consider in their legislative deliberations? This article examines four dimensions of the content of the written information state representatives in Indiana, Massachusetts, and Oregon used in 1989: whether it was policy or political information, one sided or multisided, in agreement or disagreement with the position of the legislator using it, and whether it had any hard or soft scientific content. Legislators are found to use information heavily dosed with political preferences, and they tend to look only at one side of an issue—the one with which they agree. However, they also use a substantial amount of scientific information.

Murphy, N., and S. Krimsky

2003 Implicit precaution, scientific inference, and indirect evidence: The basis for the U.S. Environmental Protection Agency's regulation of genetically modified crops. *New Genetics and Society* 22(2):127-143.

Notes:

- Literature review and analysis

• Regulation of genetically modified organisms was examined
• Decision makers were federal bureaucrats with the U.S. Environmental Protection Agency
• Conclusions: The study found that the Environmental Protection Agency used a precautionary approach when developing regulations under scientific uncertainty. The agency relied on extrapolation from limited scientific knowledge and thus presented an example of science-based policy making that was guided more by normative judgments than science.

National Oceanic Atmospheric Administration (NOAA)
2003 *Social Science Research Within NOAA: Review and Recommendations*. Washington, DC: National Oceanic Atmospheric Administration.

Notes:

• Expert social science review panel conducted interviews and literature review and analysis
• Regulatory decision making for marine and fisheries resources
• Decision makers were federal bureaucrats with NOAA
• Conclusions: The review panel presented two general findings and eight detailed findings. Overall, it reported that "Assistant Administrators were receptive to discussing the role of social science within their line offices. These discussions revealed that the full potential for social science is not being realized throughout NOAA. While social science is sometimes applied to calculate the value of scientific plans and programs, it is less often used to help identify the scope and content of science plans and programs, to evaluate the degree to which NOAA products and services are satisfying constituent needs, or to develop a more informed and participatory constituency through education and outreach programs."
• Recommendations: The review panel presented recommendations for each specific finding but in general it encouraged NOAA to focus on developing social science research priorities in two areas: (1) Programmatic: mission-driven social science research focusing on questions that provide background and operational information that will help NOAA define and effectively carry out the mandates of each line office. (2) Organizational: institutional social science research focusing on providing information related to how NOAA and each of the line offices should be organized to enhance the ability to perform required services and produce necessary outputs.

Nelson, C.E., J. Roberts, C.M. Maederer, B. Wertheimer, and B. Johnson

1987 Utilization of social science information by policymakers. American Behavioral Scientist 30(6):569-577.

Notes:

- Literature review and analysis
- Use of social science in policy decisions was examined
- Decision makers were at the federal level
- Conclusions: The study presented factors influencing use: internal sources were more likely to be used; validity and reliability were checked against one's own experience and beliefs; policy makers preferred anecdotal and soft language information over statistics; policy issues must be well defined if social science information was to have an impact on decisions; and social science information was more likely to be used if it was easily accessible and there were opportunities to clarify results and implications with the researchers. Policy makers who have a reasonable appreciation of both scientific and political aspects were more likely to use social science information. In policy making, social science information was often used in decision preparations so it provided a base on which decisions were made versus in business; such information was used directly because there was a higher cost visibility and objectives were easy to quantify, and there were also fewer constituencies to please.

Oh, C.H.

1996a Information searching in governmental bureaucracies: An integrated model. *American Review of Public Administration* 26(1):41-70.

Notes:

- Literature review and analysis, multiple regression, integrated model, and path model of information searching
- Mental health policies in service provision and financing were examined
- Decision makers were bureaucrats at federal (60), state, and local (419) levels (same dataset as Oh and Rich, 1996, article)
- Conclusions: The study summarized past research findings and those from this study, including decision makers faced with familiar unambiguous problems generally sought information within their agency—this was also true if they had a negative attitude toward social science information; however, if the problem was unusual they were likely to seek information from a wide variety of external sources; organizational norms and rules strongly influenced how information was used, for example, a decision maker's position in the organizational hierarchy determines what

information they can use and how; decision makers often used information that supported their perception of the organization's interests; decision makers' attitude toward and need for certain types of information influenced utilization; demand for information was often strongly related to its cost so this also led to utilization of internal sources; decision makers concerned about quality of methodology often turn to external sources; the more organizations had incentives for using information, the more decision makers were encouraged to seek sources outside their organization, but this varied dependent on where information processing took place in the organization.

1996b *Linking Social Science Information to Policy-Making.* (Political Economy and Public Policy Series), W. Breit and K.G. Elzinga, eds. Greenwich, CT: JAI Press.

Notes:

- Literature review and analysis, multiple regression, integrated path model, study examined mental health policies in service provision and financing
- Mental health policies in service provision and financing were examined
- Decision makers were bureaucrats at federal (60), state, and local (419) levels (same dataset as Oh and Rich, 1996, article)
- Conclusions: The study presented findings similar to the author's other articles but this one had a more detailed analysis of literature. It presented three roles of information in policy making: (1) an instrumental role in which information is used to directly influence a decision, (2) a justification role in which information is used to legitimize a set decision or the process itself, and (3) a conceptual and enlightenment role in which information is used to identify new issues and options. The following factors were found to influence social science utilization in this study: rapidly changing policy issues, organizational incentives, the decision makers' position in the organization and their attitude toward research, and information sources and types. The impact of the information was better explained by information characteristics, such as amount available or the source of it, than by other factors.

1997 Explaining the impact of policy information on policy-making. *Knowledge Policy: International Journal of Knowledge Transfer and Utilization* 10(3):25-55.

Notes:

- Literature review and analysis, statistical data analysis, integrated path model

• Mental health policies in service provision and financing were examined

• Decision makers were bureaucrats at federal (60), state, and local (419) levels (same dataset as Oh, 1996a, b, and Oh and Rich, 1996, articles)

• Conclusions: The study findings were similar to the author's other articles but this one presented a more detailed search for causal linkages among characteristics of organizations, decision makers, and information. The study found that demographic factors, such as age and education, rarely have an influence on the impact of social science information on policy making. It also found that decision makers consciously judge how much information will be helpful rather than just assuming that the information has an impact simply because they used it. Information source was found to be the most important variable in accounting for impact; decision makers were more likely to believe that information influenced the decision-making process if it came from internal sources.

Oh, C.H., and R.F. Rich

1996 Explaining use of information in public policymaking. *International Journal of Knowledge Transfer and Utilization 1996* 9(1):3-35.

Notes:

• Multiple regression, integrated path model, literature review and analysis

• Mental health policies in service provision and financing were examined

• Decision makers were bureaucrats at federal (60), state, and local (419) levels (same dataset as Oh, 1996a, b, article)

• Conclusions: The study found that information utilization was directly and indirectly influenced by a variety of factors and the links between them. Three examples included (1) Policy makers are more likely to use information in making decisions when they are faced with unfamiliar problems. In such cases they will seek a wide variety of information from a variety of sources because they need to reduce the uncertainty. (2) Information utilization was more complex in the financing area because the more technical issues and greater expertise and professional knowledge required finance decision makers to break trends—meaning that even if they had a negative attitude toward policy information, they would use it because they needed to cope with problems and persuade their colleagues. (3) Information source was the most influential factor in accounting for information use. Information from internal sources was more likely to be used, perhaps because it was easier or less expensive to obtain. The authors suggested that organizational incentive systems could facilitate wider information searches but could not guarantee information use in decision making. This can be explained with the idea that too much information from too many sources could confuse decision makers so they do not know what information to use.

Patton, M.Q., P.G. Smith, K.M. Guthrie, N.J. Brennan, B. Dickey Grench, and D.A. Blyth

1977 In search of impact: An analysis of the utilization of federal health evaluation research. In *Using Social Research in Public Policy Making*, C.H. Weiss, ed. (Policy Studies Organization Series.) Lexington, MA: Lexington Books.

Notes:

- A random sample of 20 case studies, interviews with decision makers from each case
- National health program evaluations were examined
- Decision makers were at the federal level
- Conclusions: The study found that utilization may often be defined too narrowly to include the most common uses of information in policy making. Policy makers use social science information in more subtle ways than researchers might desire. It was often used to reduce uncertainty in the decision process, such as supporting already known facts, resolving confusion or misunderstandings, improving credibility, etc. Eleven factors were analyzed for their impact on utilization: methodological appropriateness, timeliness, lateness of report, positive-negative findings, surprise of findings, central-peripheral program objectives evaluated, presence or absence of related studies, political factors, government-evaluator interactions, and resources available for the study. Two factors emerged as having significant influence on social science utilization: methodological quality and appropriateness and political factors.

Pollard, W.E.

1987 Decision making and the use of evaluation research. *American Behavioral Scientist* 30(6):661-677.

Notes:

- Literature review and analysis
- Evaluation research was examined
- Decision makers were individuals and groups
- Conclusions: The study found that evaluation research could be used descriptively for creating awareness of problems; problem definition, determining who was affected, the scope of the problem; evaluation of alternative options for solutions; consequences

involved in outcomes; and assessing implementation and effectiveness of decisions.

Powell, M.R.

1999 *Science at EPA: Information in the Regulatory Process.* Washington, DC: Resources for the Future.

Notes:

- Case study, interviews with over 100 respondents
- Use of scientific information in environmental decisions was examined
- Decision makers were federal bureaucrats in the U.S. Environmental Protection Agency (EPA)
- Conclusions: The study found that there was a weak-to-nonexistent feedback loop between decision makers and science sources; internal gatekeepers and intermediaries had strong influences on what science gets communicated to EPA decision makers; the EPA must rely on external research sources due to budget constraints so it rarely has much say in the design of the studies on which it depends; and the availability of accepted data, methods, and scope of analysis influenced what information did or did not get communicated to EPA decision makers. This study also showed the common use of case studies in environmental research.

Rich, R.F.

1977 Uses of social science information by federal bureaucrats: Knowledge for action versus knowledge for understanding. In *Using Social Research in Public Policy Making,* C.H. Weiss, ed. (Policy Studies Organization Series.) Lexington, MA: Lexington Books.

Notes:

- 38 interviews
- Continuous National Survey data were examined
- Decision makers were federal bureaucrats in seven domestic service-oriented agencies
- Conclusions: The study found that policy makers valued survey research information; they were open to developing new information utilization in their agencies; they held some feelings of mistrust toward researchers but it did not seem to prevent the use of research results; and they were aware of the needs, expectations, and constraints that researchers face but were still eager to make use of available researchers and information.

1981 *Social Science Information and Public Policy Making: The Interaction between Bureaucratic Politics and the Use of Survey Data.* San Francisco: Jossey-Bass.

Notes: Full text unavailable for review

Abstract: Published in the Jossey-Bass Social and Behavioral Science Series, with a Foreword by Kenneth Prewitt and a Preface by the author. Analyzed in seven chapters, are results of a National Science Foundation administrative experiment, the Continuous National Survey (CNS), to improve use of social science data by policy-making agencies, e.g., HEW and HUD. Interviews with CNS personnel (N = 38) over a two-year period suggested that policymakers' use of information is determined by personal or agency interests rather than by data content, cost, or timeliness. Chapter (1) Experiment in the Application of Survey Research—describes the rationale of the CNS and indicates researcher/agency communication problems. (2) Continuous National Survey: Structure and Analysis—characterizes the multipurpose nature of the survey and analyzes knowledge-inquiry systems. (3) Development and Funding of the Survey Experiment—describes the role of the National Opinion Research Center (NORC) in implementing the study and indicates priorities in determining the granting of funds. (4) Planning and Conducting the Project—describes problems arising between agencies and the NORC. (5) Assessing the Survey Experiment—points to factors of trust and agency procedure influencing data use and judges the success of the knowledge transfer mechanism. (6) Utilization of the Survey Information—suggests that data use is conditioned by involvement in collecting information. (7) Future of Survey Research for Meeting National Needs—designates bureaucratic practice as the main factor conditioning data use. four Appendixes: (A) Questionnaires; (B) Basic Coding Sheet and Summary Tables; (C) Agency Memos, I; (D) Agency Memos, II. 15 tables, references.

2001 *Social Science Information and Public Policy Making,* Second Edition. (NORC Series in Social Research: Jossey-Bass Social and Behavioral Science Series.) San Francisco: Jossey-Bass.

Notes:

- Case study analysis, 38 interviews
- Continuous National Survey data were examined (same dataset as from 1977 article)
- Decision makers were federal bureaucrats
- Conclusions: The study found that utilization was influenced by the clarity of initial definitions for specific policy applications, but information was often used for different purposes than it was initially requested for; the gap between researchers and policy makers was usually bridged easily once communication began through departmental decision-making channels; and bureaucrats sought to control information resources and processes in order to maximize the organization's interests as they perceived them.

Rich, R.F., and C.H. Oh
2000 Rationality and use of information in policy decisions—A search for alternatives. *Science Communication* 22(2):173-211.
Notes: Full text unavailable for review
Abstract: In the field of knowledge acquisition, dissemination and utilization, and impact, few studies have examined the appropriateness of rational actor theories as a theoretical framework. Rather the rational actor perspective has been simply taken for granted as a relevant analytical tool for explaining the use of information in policy making. This article singles out one major set of assumptions embedded in rational actor theories, those dealing with information acquisition and processing in individual decision making, and empirically examines to what extent the assumptions are realistic. It then puts forward an organizational interest and a communications perspective as alternative explanations for information processing in individual and organizational decision making. The findings of this article show that decision makers' behavior does not conform to the assumptions put forward by the rational actor theorists. Instead, the organizational interest perspective is far more promising in accounting for the actual behavior of individuals in processing information in making policy decisions.

Rosen, P.L.
1977 Social science and judicial policy making. In *Using Social Research in Public Policy Making*, C.H. Weiss, ed. (Policy Studies Organization Series.) Lexington, MA: Lexington Books.
Notes:
- Literature review and analysis
- Study examined the use of social science in setting legal policy
- Decision makers were Supreme Court justices
- Conclusions: The study found that it was very difficult to determine the true use of social science in legal decision making. Sometimes it was inadmissible, sometimes it was used because findings were credible but not necessarily scientific so results were partisan instead of objective; if judges wanted to change or set policy, social science information could provide the basis of empirical knowledge needed to overcome precedent; judges who are "result oriented" may be more likely to look to social science for information on potential outcomes of decisions.

Sabatier, P.
1978 The acquisition and utilization of technical information by administrative agencies. *Administrative Science Quarterly* 23:396-417.
Notes:
- Literature review, multivariate analysis

- Science and technology information was examined
- Decision makers were federal-level administrative agencies
- Conclusions: The study identified and examined variables that affected the provision of technical information including available resources, characteristics of the issue, legal and political context, and the anticipated reaction of decision makers. It also presented variables that affected the influence of technical information on decision making, including resources of information source, content of the message (here the author noted that natural scientists had greater credibility than social scientists), timeliness of the message, and resources and perspective of the decision maker. Overall, the study found that technical information was most likely influential when it involved high-quality research of a specific issue by a notable scientist who held excellent credibility with the decision maker; the findings were generally consistent with those of other studies, presented in a timely and suitable manner, and did not imply substantial changes from the decision makers' predisposed position. In addition, the influence was maximized on issues where there was high consensus on the objectives, but only moderate scientific complexity and information was most likely used in politically secure offices dominated by collegial professionals versus hierarchical managers or procedural lawyers.

Scott, R.A., and A.R. Shore

1979 *Why Sociology Does Not Apply: A Study of the Use of Sociology in Public Policy*. New York: Elsevier.

Notes:

- Literature review and analysis
- Social science research on issues of national domestic problems in the twentieth century was examined
- Decision makers were at the federal level
- Conclusions: The study reviewed past studies and found two primary factors: (1) Many applied social science studies have reported interesting findings, but few produced policy recommendations of any kind, and (2) in cases where recommendations were made, they were often rejected by federal policy makers as politically unfeasible, administratively undoable, or simply not practical. Two reasons for these outcomes were presented: (1) problems with the starting points in sociology, such as weak theory, primitive research methods, incomplete knowledge, and misperceptions on the part of sociologists of how social science research can be used by policy makers, and (2) problems with government receptivity of social science research, for example, using social science data to further agency aims, congressional members seeking social

science to reinforce preestablished positions, bureaucrats seeking it primarily to justify and refine administrative procedures and secondarily to accomplish policy, and the executive office seeking specific results to assist in developing comprehensive programs that are politically feasible. Overall, social science is most relevant to policy making as a source of methods and techniques and as providing scientific justification for one position or another. It is less relevant as a source of intellectual advice about broad policy questions or long-range implications and consequences of proposed policy alternatives. This is because the political process tends to develop policies that present a broad consensus rather than suggestions for changes that might be significant departures from the status quo.

Useem, M.

1977 Research funds and advisors: The relationship between academic social science and the federal government. In *Using Social Research in Public Policy Making,* C.H. Weiss, ed. (Policy Studies Organization Series.) Lexington, MA: Lexington Books.

Notes:

- Questionnaire survey
- A random sample of 500 academic social scientists from each of the following disciplines: anthropology, economics, political science, and psychology (1,079 usable responses) was examined
- Decision makers were academic social scientists in their role as advisors for allocation of federal research funds
- Conclusions: The study found that substantial amounts of funding were awarded to researchers whose work was valued by professional colleagues but not generally by policy makers. Advisory positions were often filled with social scientists who have greater loyalties to their academic discipline than to the federal agency involved in the funding and research.

Webber, D.J.

1987 Legislators' use of policy information. *American Behavioral Scientist* 30(5):612-631.

Notes:

- Structured interviews with a representative sample
- Social science analysis: "Policy information" was defined as scientific and technical information about the ways a policy actually works, or would work if it were to be adopted—information ranging from commonsense knowledge to academic research.
- Decision makers were state level—60 of the 100 members (65 Republicans and 35 Democrats) of the Indiana House of Representatives during the 1981 session.

• Conclusions: The study found that decision makers were not likely to use policy information or social science if left to their own inclinations. They were more likely to use these sources if they already viewed social science as valid and useful information or if their constituents requested such information, asked questions about it, or demanded that attention be paid to issues covered by it.

• Recommendations: The author suggested that academic policy researchers need to reevaluate their role as educators to focus on more interdisciplinary, decision-focused training so that students become information-seeking decision makers. Policy researchers who are interested in the use of their work must alter the knowledge dissemination process so that their research more readily becomes common sense or ordinary knowledge.

Weiss, C.H.

1977 Research for policy's sake: The enlightenment function of social research. *Policy Analysis* 3(4):531-545.

Notes:

• Literature review and analysis, 255 interviews (same data and results used in Weiss and Bucuvalas, 1977)

• Study explores use of social science for enlightenment of policy issues

• Decision makers were in federal-level mental health agencies

• Conclusions: The study presented the enlightenment model as a role for research as social criticism. It also identified characteristics decision makers used to judge information usefulness, including research quality, conformity to user expectations, action orientation, challenge to status quo, and relevance to issues the office dealt with. One unexpected finding was that decision makers were open to controversial research that made them reassess comfortable assumptions and they found it possible that others in their field would consider such research in their decision-making processes.

1979 The many meanings of research utilization. *Public Administration Review* 39(5):426-431.

Notes:

• Literature review and analysis

• The study explored the meaning of "using research." It examined six models of knowledge utilization: (1) knowledge driven—from the natural sciences, basic research reveals opportunities that may be relevantly applied to policy; (2) problem solving—direct application of results from specific social science study to a pending decision; (3) interactive—incorporates linear order from research to decision and nonlinear interconnections; (4) political—use of research to support a predetermined stand on an issue or decision; (5) tactical—strategic use of the research process or results, for example, the results may not be emphasized as much as the fact that research is being done on the issue; and (6) enlightenment—idea that social science data and generalizations permeate throughout informed publics and shape the way people think about issues.

• Conclusion: The study found that to better understand the complex ways social science is used in policy making, researchers must better define what it means to "use research."

1983 Ideology, interests, and information: The basis of policy positions. In *Ethics, the Social Sciences, and Policy Analysis,* D. Callahan and B. Jennings, eds. (The Hastings Center Series in Ethics.) New York: Plenum Press.

Notes:

• Literature review and analysis

• Social science analysis of the policy process

• Conclusion: The author argues that every policy results from interactions among ideologies, interests, and information. She presents how research is influenced by these three factors and examines the role of power in determining whose ideology, interests, and information will determine outcomes in policy making. She also suggests that in order to better understand how research will influence policy, future studies need to consider the interplay of ideologies, interests, and existing information in the situation at hand.

Weiss, C.H., and M.J. Bucuvalas

1977 The challenge of social research to decision making. In *Using Social Research in Public Policy Making,* C.H. Weiss, ed. (Policy Studies Organization Series.) Lexington, MA: Lexington Books.

Notes:

• Examined 50 research studies: conducted 255 interviews, 510 (analytic) case studies clustered through factor analysis (same data and results from the Weiss, 1977, article)

• Information use from mental health areas was examined

• Decision makers were at the federal and state level

• Conclusions: The study identified characteristics decision makers used to judge information usefulness, including research quality, conformity to user expectations, action orientation, challenge to status quo, and relevance to issues dealt with.

1980 *Social Science Research and Decision Making.* New York: Columbia University Press.

Notes:

- Literature review and analysis, 255 interviews (same data and results used in Weiss and Bucuvalas, 1977)
- Information use from mental health areas was examined
- Decision makers were at the federal and state level
- Conclusions: The study presented similar findings as Weiss's other articles and book chapter but this one had a more detailed literature analysis. Study presented the following obstacles to research use: academic researchers were often not interested in policy-relevant issues; research questions did not match policy makers definitions of problem issues; researchers simplified problems to make them easier to study; social science had few broad theories that could be applicable to framing policy research; social science methodology was often limited (i.e., data were limited or inaccurate); problems were often conceptualized to fit the methods instead of fitting the nature of the policy question; social research often took more time than policy makers had before a decision must be made; social research concepts often did not match decision makers' assumptions of social behavior; a great deal of social research examined issues that policy makers could do little to change (i.e., race, class, etc.); much of the research had inconclusive or repetitive findings or little guidance in the results; research was based on past experiences and may not match the present problems; researchers may be unwilling to make the leap required to go from data to recommendations; researchers political preferences, which influence their work, may be at odds with the perspectives of government officials; the same old social problems may not attract the interests of researchers even though decision makers still needed advice on them; results of studies in the same issue area may be divergent and contradictory. "[R]esearch is seldom used to affect decision deliberately. Rather it fills in the background, it supplies the context, form which ideas, concepts, and choices derive" (p. 155).
- Recommendation: The authors suggested that asking groups of decision makers what they want in research could be helpful for improving utilization.

Wilensky, H.L.
1997 Social science and the public agenda: Reflections on the relation of knowledge to policy in the United States and abroad. *Journal of Health Politics, Policy & Law* 22(1):1241-1265.

Notes:

- Literature analysis and persuasion
- Effects of social science research on social issues, such as crime prevention and labor market policies in the United States compared with Europe were examined
- Decision makers were at the federal level
- Conclusions: The study found that social research increased the knowledge of policy makers in two ways: (1) It helped identify issues that were open to alternatives and possible to change and (2) it brought new options and a greater range of alternatives to light.

Appendix B

Improving Environmental Decision Processes

Robin Gregory and Timothy McDaniels

INTRODUCTION

Environmental decision processes refer to the ways by which individuals, groups, and organizations—and ultimately societies—go about making choices that have implications for the natural environment. Individual decisions about consumption patterns, energy use, the relative importance of different water and air quality objectives, and trade-offs such as those among recreation options all have important environmental consequences. Choices by groups concerning alternative uses of the local resource base, including how such decisions will be made and who will participate, affect not only environmental but also economic, social, and cultural options for communities and regions. Choices by organizations and by corporations about which products to use, produce, and market; how to manage waste products; and how to incorporate learning and make investments over time all may have significant environmental implications. This includes a vast array of choices by municipal, state, national, and international agencies and governments that also have environmental significance, including regulations, land-use rules, standards for transportation and energy policies, guidelines for the extraction of renewable and nonrenewable resources, international agreements on climate change, and countless others. Considering how to go about making these decisions raises issues relating to the construction of social norms and values, cognition and emotion, analysis and discourse, information and informed consent, and the ability to appropriately frame and address difficult trade-offs. Together, these issues serve to shape how environmental decisions are now made and also set a framework for the consideration of improvements in environmental decision making as the result of future research in the social and behavioral sciences.

Sound environmental decision making reflects the theory and practice of general principles for good decision making. These include commonsense steps, such as defining the problem correctly or identifying specific objectives that will be used to assess the pros and cons of alternatives, as well as being attentive to the limits of scientific knowledge, recognizing influences of the regulatory and political context, and the like. To a large degree, this same mix of common sense and awareness of constraints would help to define good decision making in other policy areas such as health protection or space exploration or economic development.

In addition, several characteristic aspects of environmental decision making muddy the theory and complicate the application of decision making to environmental problems. One of these is the importance of scientific knowledge to making good decisions. A second is that the needed scientific knowledge encompasses both the natural sciences and the social sciences. It is well known that the quality of communication between these two groups of professionals is often poor. A third characteristic of many environmental decisions is the level of uncertainty and disagreement associated with the science. Finally, environmental decision making often incorporates scientific and expert understanding and analysis within an explicitly deliberative and political context (National Research Council, 1996) in which technical experts work alongside community residents, representatives of local or state governments, and consultants or members of nongovernmental organizations and other interest groups. As a result, there is a need to combine the knowledge of scientists with that of local residents and resource users in ways that are readily understood by diverse groups of stakeholders and to use processes that help these same groups to make good decisions.

In this appendix, we link two primary fields of study, one based in the decision sciences and the other in environmental policy, and draw from a variety of other disciplines, including psychology, economics, political science, and geography. Rather than offering a comprehensive review of these contributions, we identify some key insights and promising areas of research that can be useful for improving decisions made by individuals, groups, and organizations that may affect the natural environment. We emphasize topics that have the potential for improving environmental decisions within the next decade. In each of these topic areas, substantial progress already has been made, but much more can be done.

LEVELS OF ENVIRONMENTAL POLICY DECISIONS

Environmental decisions include management strategies or levels of funding for activities that either directly affect components of the natural environment, such as forests, oceans, wildlife, inland waterways, or fisheries, or influence health and lifestyle issues by, for instance, managing toxic wastes or air or water pollution. Yet the scope of environmental decisions is far broader because of the relationships among environmental management choices and economic, social, health, and cultural values. Environmental choices influence local and national economies, the geographic distribution of employment opportunities, people's health and longevity, the social structure of communities, and, to an underappreciated degree, the economic and social fabric of the country. Moreover, economic, social, and health choices hold important implications for management of the environment. Many of the environmental challenges facing society, ranging from climate change to species diversity to genetically modified crops to soil contamination, stem directly or indirectly from decisions made primarily on the basis of economic, social, and political concerns rather than environmental considerations.

Environmental decisions extend across time scales, physical scales, and institutional scales. Although it is commonplace to observe that issues of environmental sustainability have dimensions that extend from the local to the global, relatively little of the writing on environmental decision making actually addresses how such linkages should be addressed in deliberation, assessment, analysis, and management (Cash and Moser, 2000). Good environmental decision making requires processes that link and balance values and technical information about impacts across these multiple scales, over long time horizons. This is not an easy task.

Beyond the complexity introduced by multiple scales, environmental decisions remain among the most difficult, multifaceted, multidisciplinary questions facing society. They are fraught with abundant uncertainties, value conflicts, long time horizons, high stakes, severe organizational and institutional constraints, and many deep levels of emotion.

We have organized the discussion around who it is that faces the decision and makes a choice: individuals, groups, and organizations. As noted in the following discussion, we acknowledge that some issues (e.g., adaptive management policies as a concern for organizations) fit quite neatly into one category whereas other issues (e.g., difficulties in making trade-offs, the role of time) cut across all three levels.

Research and Practice on Environmental Policy Choices by Individuals

Much of the research conducted by decision scientists over the past 50 years has addressed questions of individual decision making from two primary perspectives. One is normative, as reflected in the domain of subjective expected utility theory, which is built on a set of rational axioms that show the conditions for optimal choices in light of uncertainties and multiple objectives (Keeney and Raiffa, 1993). Subjective expected utility theory is widely recognized as applying to idealized decision making because it says very little about "how to do it": that is, the processes by which these concepts could best be implemented in practice. The second perspective is descriptive, examining how individuals actually make choices in the face of complexity, limited time and information, and the need to balance concerns for accuracy and effort (Kahneman, Slovic, and Tversky, 1982; Payne, Bettman, and Johnson, 1992). Descriptive research has provided valuable insights into unaided human processes of perception, judgment, and choice, but says little about how one should design or conduct decision processes so as to make better choices. A third perspective, of prescriptive decision aiding (Bell, Raiffa, and Tversky, 1988), has received far less attention despite the obvious importance of helping people to make better decisions and choices and the robust research finding that, when left to their own devices, people "systematically violate the principles of rational decision making" (Slovic, Fischhoff, and Lichtenstein, 1976:169).

Individuals facing environmental choices need to first determine what is being asked of them and then figure out their preferred response, communicating it in an appropriate manner (to ensure that their "vote" will count and be entered correctly). This involves identifying their values in terms of those concerns (environmental, economic, social, health) that might be affected by the reasonable set of actions or policy choices. Researchers in the decision sciences have emphasized that this process generally occurs in the context of significant uncertainty, concerning both what the individual might want (Slovic, 1995) and what the consequences of different decisions might be (Morgan and Henrion, 1990).

Behavioral decision research, in particular reflecting the descriptive findings by Kahneman and Tversky (e.g., Tversky and Kahneman, 1981), has emphasized that a variety of heuristics and biases, such as anchoring on first impressions (without sufficient later adjustment) or paying undue attention to more salient aspects of a choice, may systematically influence the decisions made by individuals, with the result that choices may differ from those that these same individuals might, upon reflection, prefer (Kahneman et al., 1982). Findings by decision scientists (principally cognitive psychologists and behavioral economists) also introduce the idea that preferences for many environmental, social, and economic choices are constructed rather than simply revealed by a measurement practice (Slovic, 1995). If people are familiar with a choice and know what they want (e.g., because they have learned through repeated trials), then it might not matter much how they are asked. But when faced with unfamiliar decisions or trade-offs across novel options (neither of which is uncommon in the realm of environmental policy choices), individuals may not know their own preferences and therefore are required to construct (as well as articulate) their values in the context of a specific choice. This will be accomplished by building from an underlying set of more basic values, guided in part by the cues provided by the elicitor or the judgmental setting. The reality of constructed preferences means that the analyst must ensure that such cues come from relevant sources (e.g., experiences with similar types of goods) rather than irrelevant ones (e.g., unintentional nonverbal hints).

The construction of preferences and, in turn, observed choices and judgments will be influenced by affective and emotional considerations as well as cognitive processes. Research by neurologists such as Damasio (1994) and decision scientists such as Loewenstein (1996) demonstrates that a key predictor of a person's valuation of an item will be their general assessment of the positive or negative affect associated with the good. Affective considerations that come into play as part of a decision-making process, reflecting the individual and the task as well as the interaction between them, influence relative judgments of salience that, in turn, influence multidimensional evaluations of options (Slovic, Finucane, Peters, and MacGregor, 2002). Mellers, Richards, and Birnbaum (1992), for example, showed that the weights in integrative valuation tasks are inversely proportional to their variance, a result of the greater affective impression on judgments made by attributes whose variance is smaller.

Research and Practice on Environmental Policy Choices by Groups

Deliberative processes that seek to obtain responsible input to environmental decisions by involving small (10 to 25-person) groups in discussions about the choice of a preferred environmental policy have become very popular over the past 15 years. Guidance has been provided by a variety of official and quasi-official publications and bodies, including the National Research Council (1996) and the Presidential-Congressional Commission on Risk (1997). Similar initiatives have been undertaken in Canada (Canadian Standards Association, 1997), the UK (Royal Commission on Environmental Protection, 1998, and other countries (e.g., Australia, The Netherlands, Germany).

Successful deliberation requires a combination of at least three elements (Gregory, Fischoff, and McDaniels, 2004). The first is agreement among participants on the ground rules for participation; this involves both bringing the set of legitimate stakeholders to the table and keeping them there with appropriate rules for dialogue, analysis, and addressing disputes. The second element is a process for aiding decision making by group members that provides a context for creating effective understanding. Third, successful deliberation requires techniques for integrating the views of participants, for translating opinions into values, and for communicating effectively with decision makers. Addressing these three elements effectively requires an understanding of both analytical components, in terms of distinguishing technical (factual) and value-based issues and explicitly addressing sources of uncertainty (relating either to facts or values) as part of selected evaluation approaches, and behavioral components, in terms of helping participants to understand the issues and express their conclusions clearly.

A wide variety of formats, principles, methods, and techniques have been proposed, employed, and analyzed to varying degrees as a basis for conducting deliberative processes. Renn, Webler, and Wiedemann (1995), for example, discuss approaches ranging from citizen juries to stakeholder negotiations. Beierle and Cayford (2002) and Chess and Purcell (1999) discuss the evaluation of such processes. The range of potential models is vast and, as discussed in more detail below, few empirical comparisons of different approaches have been conducted.

Our experience is that the methods of decision analysis, which typically have been applied to individual choices (Hammond, Keeney, and Raiffa, 1999), provide a particularly useful model for guiding group deliberations that address public policy questions. The distinguishing feature of decision analytic approaches is that they are sensitive to human judgment. People who might be influenced by a decision are asked to develop a clear statement of their relevant values, in terms of what matters in the context of the decision alternatives and operative constraints. Technical experts are asked to provide information about the consequences of these options as well as the associated uncertainty (e.g., probability distributions, degrees of belief); individuals are then asked to weigh the various objectives in terms of their relative importance, in the context of the specific problem under consideration. Both computational and judgmental methods are used to combine these components, using the precepts of multiattribute utility theory (Keeney and Raiffa, 1993) to summarize across concerns and provide recommendations to policy makers.

Research and Practice on Environmental Policy Choices by Organizations

Although behavioral decision theory has focused on the individual, helpful insights have been gained into the behavior of organizations and the ways in which they make decisions and address multidimensional choices (March, 1997). Much of this work is descriptive, with substantial progress made in recent years on understanding how organizations change in response to new ideas and stimuli. Social learning theory, which identifies ways in which organizations transform knowledge into action, has been one of the main themes in research on organizational theory over the past two decades (Levitt and March, 1988; Argote, 1999).

A widely discussed application of social learning in the context of environmental decision making is adaptive management (Holling, 1978; Lee, 1993). The premise is simple: because there are profound uncertainties in resource management decisions, policy actions should be regarded as experiments and a positive value should be attached to characteristics of management plans such as flexibility, learning, and monitoring or feedback. The idea of adaptive management has common-sense appeal, and the underlying concept of science-based policies as testing hypotheses and embracing failures is basic to the scientific method. Yet significant institutional, political, and organizational barriers have blocked many of the more innovative plans to pursue adaptive management strategies (Gunderson, Holling, and Light, 1995).

Another important concern for institutions, in the both private and public sectors, involves consideration of the costs and benefits that stem from actions (projects, programs, or policies) that occur at different points in time. To facilitate these intertemporal comparisons, organizations typically calculate the present value of future gains and future losses; on the basis of prescriptions that follow from the expected utility model, the usual practice is to employ a single, invariant rate of discount. This practice, however, is not supported by behavioral studies; as discussed further in the section "Approaches to Aid Organizational Decisions on Environmental Policy Changes" below, the topic is one of many areas receiving significant attention from researchers.

APPROACHES TO AID INDIVIDUAL DECISIONS ON ENVIRONMENTAL POLICY CHOICES

We noted above that the field of decision sciences has benefited from several syntheses that provide solid assessments of the field. Influential books were written by Kleindorfer, Kunreuther, and Schoemaker (1993), who provide a broad review of the field of decision sciences at the individual, group, and societal levels, and by Payne et al. (1992), who address individual decision making from the viewpoint of behavioral decision research. Mellers, Schwartz, and Cooke (1998) focused their review on behavioral violations of rational choice theory; Pidgeon and Gregory (2004) summarized contributions of the decision sciences to public policy applications, including a review of the important role played by heuristics as a means for cognitively simplifying otherwise dauntingly complex choices faced by individuals. In this section we provide an overview of some of the key concepts from the decision sciences related to individual preferences and preference elicitation. After that we turn to a number of topics that could help comprise future research priorities related to these issues.

We noted above that people often do not hold well-defined values for complex and unfamiliar environmental goods. Hence, understanding preferences should stress the important role of helping people *construct* their preferences rather than simply revealing them through actions or questions (Slovic, 1995). Tools concerned with preference construction and elicitation include formal methods to elicit preferences, based on precepts of measurement theory and decision theory, along with wisdom gleaned from applied experience. Perhaps the most well known of these formal approaches, multiattribute trade-off (or MAUT) analysis typically involves an interview between an analyst and a decision maker to construct a utility or value function for the decision maker. Keeney (1992) provides many examples of these kinds of questions, and the functional forms and the assumptions involved in such efforts, as well as references to many examples in which these methods have been employed. The advantages of a MAUT approach are that the judgments involved are made explicit, the value information can be used in many ways to help clarify the decision process, and the decision maker typically learns a great deal through these joint efforts to construct their views on preferences. The disadvantages can also be substantial: the questions involved may be difficult to answer and require decision makers to make their inchoate feelings explicit, how the results will be used may not be transparent, and the process can be cognitively and analytically demanding.

In part as a response to these drawbacks, and in part reflecting the richness of decision research, several other approaches to preference elicitation have been developed and tested by researchers. For example, the analytic hierarchy process, developed by Saaty (1991) is a widely applied approach to eliciting preferences in decisions with multiple objectives. Proponents maintain that it involves more transparent methods and questions than those required for MAUT analysis; critics question whether results of an analytic hierarchy process might violate normative principles of decision making. Other, more intuitive approaches to judging preferences also are found in the behavioral decision-making literature. These include making decisions using only one of the objectives, even though several are important (lexicographic ordering), or selecting the first alternative that provides acceptable performance across all the objectives "satisficing" (Payne et al., 1992). Of course, these behaviorally straightforward methods also may encourage some of the heuristics, biases, and shortcuts that can undermine the quality of decision making in ways discussed above.

A trend to simpler and less demanding methods, which make use of differences among the alternatives to help clarify preferences, is evident in recent studies on decision making. Hammond et al. (1999), for example, discuss an approach to constructing and clarifying preferences called "even swaps." This approach does not involve developing a utility or value function, but instead develops an objectives-by-alternatives, or "consequences," table based on the judgments of the decision maker to clarify the relative importance of differences in how well the different alternatives achieve the objectives for a decision. The approach then uses that insight as a basis for eliminating dominated alternatives and for expressing different objectives in common units, which in turn greatly facilitates judgments of which alternative is preferred.

Another useful analytical approach that has been developed in applied decision analysis practice is termed "value-focused thinking" (Keeney, 1992) The deceptively simple but fundamental notion is that attention to values can serve as the basis for several key steps in designing decision processes. For example, clarity about what matters and how it is to be measured is important in several ways: defining the information needed to characterize the consequences of alternatives; designing better, more widely supported alternatives; and indicating creative opportunities to improve the range of choices available (Keeney, 1992).

Identifying Subjective Judgments and Values

Scientific investigations often are viewed as "objective" in contrast to the "subjective" perspective of those concerned with social impacts, such as fear or worry, or process considerations, such as trust or fairness. Yet even the most highly complex scientific choices rest on subjective decisions that reflect what data to include, what people to ask questions of, and what methods should be used. Thus, both perspectives share a similar qualitative foundation. Choices about what people are included, what views are seen as data, and what criteria to use in analysis always are the results of judgments. Extensive research has been conducted on this topic (e.g., Jasanoff, 2002), but the audience of academics and resource managers remains largely uninformed, in part due to poor communication but also because much of the research fails to address relevant decision contexts.

Hence, research that more squarely addresses the role of subjective judgments in environmental decision making, both for issue-based topics and general policy analyses, could potentially make substantial contributions. Relatively little attention seems to have been paid to topics of problem choice or to the relationships among broad strategic objectives (such as long-term environmental, social, and economic sustainability) and near-term, more prosaic decisions like the choice of transportation modes or infrastructure options. Issues regarding the right level at which to conduct research and policy analysis should matter in shaping the kinds of analysis that are done and, ultimately, the policy decisions that are made. Other kinds of key judgments, beyond those listed above, include issues such as how we conduct analyses for linked decisions, in which current choices provide opportunities for learning over time (Keeney and McDaniels, 2002), or situations in which regulatory issues are linked across multiple scales and levels of decision making (McDaniels and Gregory, 2004). Other kinds of subjective judgments that merit research could include generalizations of the conditions on which cooperation is likely to occur among parties in commons contexts, particularly in situations where multiple levels of cooperation may be needed.

Clarifying Performance Measures

Despite the obvious need to specify project or action consequences, insufficient attention has been given to the design of performance measures (Keeney, 1992). In part, this is because of the emphasis of many policy analysts on economic methods and the model of cost-benefit analysis, which uses dollars as a common metric for evaluating impacts. Yet measures such as dollars represent only one of three primary types of indicators or attributes: natural, proxy, and constructed.

Natural attributes are in general use and have a common interpretation. The management objective to "maximize profits" is naturally measured in dollars; similarly, if one management objective is to minimize the loss of wildlife habitat, then a natural indicator might be "acres of lost habitat." Cost (also measured in dollars) and worker injuries (measured in numbers) are other examples of natural indicators.

Proxy attributes also are in general use and are well understood. However, they are less informative than natural attributes because they only indirectly indicate the achievement of an objective. An example is the use of a measure such as "returned items per $10,000 sales" as a proxy for product quality. Another common example in environmental contexts is the use of an easily measured indicator (such as air emissions in parts per million) as a proxy for impacts of concern that are harder to measure (such as adverse health or visibility impacts due to air quality impairment).

Constructed attributes are used when no suitable natural attributes exist. An example is a scale to measure community support for forest management practices. Because no natural scale exists to measure public support, an index (e.g., 1-5 or 1-10) needs to be created, with each rating denoting a different level of support. Many such constructed scales are in widespread use: the gross national product is a constructed measure, as is the Dow Jones stock average in the United States or the Apgar score given to track the health of a newborn. When thoughtfully designed, constructed indices can greatly facilitate management by defining precisely the focus of attention and by permitting trade-offs across different levels of the concern and other attributes (e.g., is it worth postponing harvest of an area for x years in order to increase support from level 2 to level 4?).

Attributes are made operational through the development of scales.

These scales serve two major purposes. First, they provide a means for distinguishing among different levels of impact with respect to the attribute. Second, the scales provide a clear means for distinguishing the endpoints of the range of anticipated impacts. As an example, consider a scale denoting the expected cost of a range of management options. If the lowest reasonable cost for the options under consideration is $20,000 and the highest reasonable cost is $70,000, then the scale should reflect this range and be measured in thousands of dollars (20-70). Using a measure such as "hundreds of dollars" is not appropriate because it conveys an unnecessary (and probably illusory) sense of precision. Similarly, converting a natural scale of this type to an index (e.g., whereby a "1" = $20,000-30,000, "2" = $30,000-40,000, etc.) is not a good idea because information is lost (i.e., is a "2 " at the high or low end of the range?). On the other hand, a scale for measuring community support for environmental plans would be a constructed scale and it might be measured in terms of several related attributes (e.g., turnout at meetings, support conveyed through a survey, etc.) so that a "1" might denote a low measure of support (low turnout, low percentage of support in a survey) and a "5" denote a high measure of support. In general, scoring methods used to select scales should be accurate, understandable, and at the appropriate level of discrimination.

In our opinion, far more attention needs to be given to the design of clear performance measures and to their incorporation as part of environmental evaluations. This increased use of performance measures would yield three primary benefits, all relating to the provision of information that will aid stakeholders and decision makers. The first is to focus attention on those aspects of the problem that are considered to be important and to set up measurable criteria by which progress on these considerations can be assessed. For example, one objective of sustainable forest management may be to maintain overall forest health. To be useful for management purposes, forest health will need to be disaggregated into components that can be identified and measured. Resilience might be one such component, and productivity might be another; if so, then measures for resilience and for productivity will need to be developed along with some criteria for weighing their relative contributions (alongside other components) to overall forest health. A second benefit would be to discriminate more clearly among competing hypotheses, therefore contributing to the conduct of scientific investigation. A third benefit would be to stimulate the identification and creation of a range of management alternatives and to serve as decision aids for the identification of a preferred management plan (or set of plans). The best management actions are those that best achieve the objectives of the environmental management problem; without clear measures, there cannot be clear communication about the ability of actions to satisfy the identified objectives.

Evaluating Trade-Offs

There are three main reasons why trade-offs are addressed poorly as part of many environmental management plans (Gregory, 2002). First, addressing trade-offs requires techniques that help the individual to explicitly address multiple dimensions of value. This task is cognitively difficult and in many cases requires the adoption of decision-aiding methods for weighing the importance of different components of the problem that are unfamiliar to many analysts and decision makers, let alone many members of the lay public. Second, trade-offs can be emotionally difficult: they raise moral and ethical dilemmas and can require individuals to address explicit choices about topics that they (and other people, including their elected leaders) may find uncomfortable (Baron and Spranca, 1997; Fiske and Tetlock, 1997). As a result, in many cases these decisions are either treated informally (to decrease perceived responsibility costs) or left to others. Third, it is often assumed that addressing trade-offs in a rigorous and defensible manner will prove cumbersome and expensive. This assumption is not necessarily true: relatively simple and straightforward techniques exist for helping people to address trade-offs in ways that can substantially assist many individual decisions and provide essential insights for negotiations between individuals whose values may differ (Hammond et al., 1999). Furthermore, even extremely difficult or complex problems often can benefit substantially from the insight provided simply by clarifying the nature of the trade-offs and how they influence choices across management options.

Using Expert Judgment Processes to Understand Uncertainty

Input from technical experts is required to help anticipate how actions might affect the natural and the human environment and to develop approaches for mitigating potentially adverse impacts. Both the identification of impact categories and the assignment of probabilities are judgmental actions, often requiring skills of inference and prediction for which most scientists (as well as most laypersons) have received little or no formal training. Current research on probabilistic analysis is highlighting ways in which the basis for technical judgments can be clarified and likelihood estimates of impact magnitude and severity can be refined and communicated (Cullen and Small, 2004).

New tools and approaches for clarifying uncertainty are intended to improve the quality of environmental management decisions over time and to increase the understanding of how a variety of techniques (such as adaptive management trials) might be used by managers as a possible response to uncertainty. At present, uncertainty often is handled in a more casual manner, which can result in suboptimal decisions and fewer opportunities for learning (because feedback in terms of management responses is more difficult to incorporate). At minimum, environmental managers should consider the different sources of uncertainty—including uncertainty about impact severity, habitat responsiveness, the influence of slow variables such as climate change, the effectiveness of mitigation measures, monitoring results, and the compliance of proponents—and decide (and document) how each source is best handled.

A variety of techniques are available to incorporate judgments of uncertainty. As one example, expert judgment techniques (Keeney and von Winterfeldt, 1991) can be employed in cases in which higher or more consequential risks are anticipated and uncertainty is high. These techniques seek to first decompose a more complex problem into its parts (which allows for simpler judgments) and then to improve the uncertainty assessments of managers for those aspects of the problem where they are either least confident (i.e., about their assessments of expected impact distributions) or for which disagreements exist among managers or across stakeholders (Morgan and Henrion, 1990). Different approaches can be employed, depending on the needs of the problem domain; as one example, it may be more appropriate to elicit degrees of belief in specified endpoints (using questions along the lines of "Given mitigation plan *X*, assign 100 points among the three future habitat states of *A*, *B*, and *C*") rather than probability distributions. Yet although formal expert judgment techniques have been used to enhance understanding of a variety of resource management problems, many natural scientists are hesitant to make use of the technique due to concerns that (a) it will highlight disagreements between individuals (whereas in fact explicit expert judgments typically lead to more agreement because the reasons for differences among views are clarified), (b) it will substitute elicitations for field trials (whereas in fact the two approaches are highly complementary), or (c) it will undermine confidence in the assessments of experts (whereas in fact the explicit documentation of uncertainty often eases worries and increases the credibility of scientific assessments) (Gregory and Failing, 2002).

APPROACHES TO AID GROUP DELIBERATIVE PROCESSES ON ENVIRONMENTAL POLICY CHOICES

The science underlying our knowledge of approaches to aid group deliberative processes for environmental choices has benefited from recent reviews. A useful starting point is again Kleindorfer et al. (1993) who provide an overview of small group decision-making processes. Renn et al. (1995) evaluate the ability of decision science methods to encourage fairness and competence in public participation. Beierle and Cayford (2002) review the content and outcomes of a large set of applied cases, providing conclusions on the benefits and drawbacks of such processes.

Criteria for the Conduct of Deliberative Processes

Deliberative processes, involving either small groups of experts (science and lay representatives) or larger panels (advisory councils or specialty forums), are used increasingly to open up the environmental decision-making process and to ensure that a wide range of views is heard. Yet, in most cases, the weighing or balancing of conflicting objectives, which is the essence of clarifying trade-offs, is either ignored or only partially addressed. Instead, there is usually some attempt for public values to be expressed (e.g., in the form of goals or concerns), but rarely is attention given to carefully structuring the underlying choices and explicitly addressing the key trade-offs. As an example, Duram and Brown (1999:462) reported after extensive interviews with stakeholders active in U.S. watershed planning initiatives that "fewer than 50% noted that participation was useful in clarifying the issues," although this would seem to be a minimum requirement of any policy-based deliberative process. The underlying behavioral questions—how well the deliberative process is understood, how thoughtfully the input is provided, and how meaningfully outputs are considered (in terms of process and outcome linkages)—remain areas of frustration to theorists and practitioners alike.

One obvious goal is to obtain judgments that are better matched to the decision at hand and more cognitively manageable by the participants. An example is the use of voting as a basis for preference elicitation in specific policy decisions. McDaniels and Thomas (1999) discuss an approach in which voters are asked to choose among a set of possible policy alternatives to address a given environmental question. This approach relies on problem structuring tools from decision analysis to develop a well-structured decision with explicit policy objectives and alternatives. Then risk communication methods are used to help voters understand the consequences of the alternatives. Voters then must make holistic judgments that integrate across the various objectives to select the preferred alternative(s).

Beierle and Cayford (2002) identify generic performance criteria for deliberative processes, emphasizing procedural matters relating to how the process is conducted. Writings by Renn et al. (1995) also emphasize issues regarding how such processes are conducted to build trust, avoid power imbalances, and foster agreement. Still other criteria, which deserve more attention, address how a given approach handles the complexity that is a crucial part of all important environmental decisions, not simply in terms of information sharing, but also how the problem structuring concepts, formal analysis of judgments, and analytical tools of decision sciences are used. Other criteria would need to take account of the role that emotion and affective considerations, as well as deliberation, play in the conduct and quality of discourse pursued by a group. Key research needs, therefore, include methods that could help to ensure that participants understand a problem (e.g., recognize that its representation is both complete and comprehensible) and that they are able to make sense of their assigned evaluative task (e.g., by using valuation methods that are cognitively appropriate and provide for the expression of affective and emotional concerns).

Linking Local Knowledge and Scientific Expertise

Many environmental policy initiatives fall short of expectations: deliberations are not perceived as open, the scientific basis for decisions is questioned, and managers are unprepared for (the inevitable) ecological and political surprises. A particular problem for environmental managers is that, in many cases, lay and community participants feel disenfranchised and believe that key elements are missing from recommended management alternatives because their concerns and values have not been heard. As a result, policy recommendations fail to reflect the full set of relevant objectives and, not surprisingly, community support is withdrawn.

One explanation for this failing (and, consequently, for the dissatisfaction of many community residents) is the apparent choice, by the initiating agency or facilitators and analysts, to emphasize science literacy and thereby place primary emphasis on the opinions of scientists and other technically trained participants (Norton and Steinemann, 2001). Better science, it is concluded, will lead to better deliberative processes. The U.S. Environmental Protection Agency's Science Advisory Board, for example, recently concluded that "stakeholder decision processes . . . frequently do not do an adequate job of addressing and dealing with relevant science" (Environmental Protection Agency, 2001). The rationale is that these are complicated technical issues and laypersons simply are not sufficiently well informed to make rational decisions. Examples include the hesitancy of many scientists to provide accurate information to public stakeholders about the uncertainties associated with risk management options due to concerns that disclosure might "cause panic and confusion regarding the extent and impact of a particular hazard" (Frewer et al., 2003). Greater reliance on science and on the judgments of scientists, it is believed, will make for better choices.

Although we agree that sound science is necessary for environmental decisions, it is often not sufficient: in particular, we believe that increased attention should be given to the significant body of knowledge held by local and community participants that is not grounded in conventional scientific methods but is nonetheless empirically derived. Some of these knowledge holders are long-time community residents; some are aboriginal populations with special interests and cultural uses of environmental resources; some are resource users with specialized knowledge such as fishers or trappers. This alternative knowledge base may be based on different techniques, and reflect differently constructed forms of knowledge, than those of Western scientific methodologies, yet we believe it often represents a useful—in many cases essential—complement to science-based knowledge. For example, in the restoration of severely disrupted ecosystems (perhaps due to construction of a hydroelectric dam), traditional and local knowledge often provides the only record of ecological processes prior to disruption, and thus provides a template for the end goals of a postrestoration landscape. In other cases, local and traditional knowledge raises concerns (e.g., about protection of a plant or animal species) that are missing from scientific analyses or highlight considerations that have not been fully examined. Alternative knowledge sources can also provide an important test of convergent validity, as for example when anthropologists and archeologists check oral history against artifacts to understand more about how people first came to populate the Americas (Glavin, 2000).

Distinguishing Process and Outcomes

The quality of environmentally significant decisions cannot appropriately be judged solely by their outcomes because those outcomes are highly dependent on factors that are uncertain and often unknown when the decision is made. One can make a decision that is well informed and well considered given what is known, but be unlucky regarding the outcome of some key uncertainties, so that the results are not as hoped. Thus, it is necessary to have explicit criteria for judging the quality of decisions. Across many decisions, higher-quality decisions (those based on the best available information, careful identification of objectives and measures, and so forth) are more likely than lower-quality decisions to lead to desired outcomes (von Winterfeldt and Edwards, 1986). It is also reasonable to expect that higher-quality decisions will be more defensible, which is of concern to many decision makers facing scrutiny from the public, other constituents, or the courts.

The problem of defining decision quality for practical environmental decisions has not received the level of research attention it deserves. Both the normative and the empirical traditions in decision science have difficulty with this problem. Normative decision theory yields prescriptions that are difficult, if not impossible, to implement in practice because of the complexity of the value judgments involved and because of the requirement to assess every consequence of every possible alternative and to evaluate them all against the values of each decision participant. Decision makers do not seek the normatively best practice, but the best practice under constraints of real people's cognitive capabilities, legal requirements, limited time and resources, social conflict, and so forth. Researchers in the empirical tradition can be uncomfortable defining decision quality because they see a conflict between normative and positive science or because they question whether any single standard can hold up in a diverse society.

The kinds of research that could be conducted might address issues as diverse as the following: How do we judge the notion of decision quality in environmental contexts? How do we provide ongoing heuristics and routines that could make the elements of good decision process more accessible and readily applied? How can decision aids (such as, for example, a CO_2 emissions calculator) be developed to help provide a structure and key information for everyday environmental choices? In addition, little is known about whether stakeholders might respond more favorably to management initiatives if different decision processes were followed; for example, could an improved process help to reduce criticism and increase acceptance in the aftermath of a low-probability but high-consequence accident such as a spill or collision? Research also has shown that reliance on, and citations of, broad stakeholder-based input can make the results of environmental decision processes more acceptable to others who are less familiar with the issues (Arvai, 2003).

APPROACHES TO AID ORGANIZATIONAL DECISIONS ON ENVIRONMENTAL POLICY CHOICES

Basic elements of research and practice regarding organizational decision making are surveyed in the writing of March (1997). Other writers, such as Argote (1999), have emphasized issues of organizational learning. There is an active field of research on how organizations respond to environmental regulations and how they structure and conduct their compliance efforts (e.g., Jennings, Zandbergen, and Martins, 2001). Policy analysis as an approach to environmental decision making in government organizations is extensively discussed in a number of textbooks, such as Weimer and Vining (1999).

The Choice of a Policy Evaluation Approach

The dominant method for the analysis of environmental options is to employ welfare economics-based cost-benefit techniques. Although a cost-benefit approach can encompass many facets of a natural resource initiative, it ultimately collapses these into a measure denoted in dollars. This has the advantage of providing a single measure with which decision makers can compare options but it also masks the contribution of individual factors, thereby making it difficult to track information about different effects (environmental, economic, social) and alienating individuals whose concerns (such as health and safety, biodiversity, or community image) might be difficult to translate into dollar measures. In addition, other approaches that make use of multiple metrics to assess the different dimensions of the problem and explicitly examine trade-offs, such as those based in decision analysis and multicriteria methods or other more qualitative approaches, tend to receive less attention.

Research questions concern the further refinement of these alternative approaches as well as the introduction of more case findings and comparative studies that can help to illustrate similarities and differences, both among techniques and in terms of how they are perceived by different groups within society. Early work by decision scientists (e.g., MacGregor and Slovic, 1986) offered interesting observations about the acceptability of different policy evaluation frameworks, but little further research has been conducted. An interesting new perspective is provided by the interest in narrative as an alternative to more analytic evaluation approaches; Satterfield, Slovic, and Gregory (2000), for example, compared the responses of participants with narrative and cost-benefit presentations of a complex environmental policy decision that required trade-offs across hydroelectric power production and fisheries health. Their results demonstrated advantages of narrative approaches for engaging participants and helping them to assess the relevance of technical information.

The Role of Learning

A crucial step in fostering good organizational response to complexity in environmental decisions is the notion of learning over time. Parson and Clark (1995) reviewed the literature on social learning (in this case, for sustainability issues), which is a solid starting point for understanding learning in organizations. This broader topic of organizational learning was surveyed by Argyris and Schon (1978) and, more recently, by Argote (1999). Policy analysts such as Lee (1993) and ecologists such as Gunderson et al. (1995) have stressed the organizational implications of adaptive management options, which involve viewing policies as experiments (discussed further in the next subsection). These concepts, which turn one-time decisions into repeated decisions with opportunities to learn and adapt, offer a wealth of new ideas for improved organizational practice in the context of complex environmental decisions.

Organizations do best in situations in which they can rely on standard procedures to address ongoing management issues. Hence, finding ways to turn complex choices into learning opportunities, and building ways in which this learning can serve as the basis for improved decision processes over time, provide important research opportunities. McDaniels and Gregory (2004) discuss the benefits to organizations of treating learning as a specific objective in decision processes with stakeholders, including the potential for developing institutional processes that foster and measure organizational learning opportunities.

Implementation of Adaptive Management Methods

Adaptive management methods are designed to help reduce uncertainty through the conduct and comparison of selected experimental trials. Although scientific support for adaptive trials is high, few examples of successful implementation can be found. The reasons have to do with the design of the trials, the evaluation of their benefits and costs, and the institutional framework within which they have been proposed. New ideas are being proposed in each of these areas; together, they can help to facilitate learning, give meaning to guiding concepts (such as sustainability and the precautionary principle) whose vagueness has led to confused implementation, and encourage the selection of improved alternatives (Gunderson and Holling, 2002).

The concept of adaptive management was born out of the need to address the objective of learning about managed environmental systems over time. Learning is most important when uncertainties are high and when management actions are (at least in part) unfamiliar and of high consequence. Prescriptively, an adaptive management approach requires four primary elements (Walters, 1986):

1. bounding of the management problem in terms of objectives
2. characterizing existing technical knowledge about the system
3. designing management treatments (involving either passive or active management)
4. incorporating measures for reducing catastrophic risk and improving long-term outcomes

Traditional monitoring of environmental management initiatives tends to be relatively cheap; adaptive management tends to be relatively expensive. But in either case, the analysis of the pros and cons leads to subsequent questions: Why do we want to monitor? What do we hope to learn? How will we know when we have learned it enough to do something different? How do we know if this is the best way to learn? At the broadest level the question is: Why do we want to do adaptive management trials? Once these questions are answered, we must decide whether these are passive or active adaptive management efforts, over what time periods these issues are to be explored, and how those time periods correspond to the time periods of change in other variables (for example, those with slower rates of change). Finally, we must consider how to structure the underlying expert judgment tasks. These are tough tasks, but solvable, in terms of being amenable to analytical techniques.

The toughest question of all, in many cases, is getting approval for adaptive management (or even for comprehensive monitoring). The only reason to do either is to know more later than is known now, in which case institutions need to be sufficiently flexible to acknowledge this learning and to do something different (and presumably better) in the future. Building in this sensible institutional response to adaptive management is not easy (for evidence, look to management of the Columbia River system over the past decades).

Reevaluating the Role of Time

All environmental decisions involve the element of time, yet it is rarely taken into account in directly meaningful ways. Intertemporal aspects of decisions clearly have to do with the occurrence of effects or consequences. Yet they also have to do with how impacts will be perceived (in terms of the values in place at the time) and how these perceptions will be coded (in terms of issues such as adaptation and vulnerability). Environmental activities such as the restoration of damaged riverine ecosystems may build on an extensive base of natural science research that takes place over time periods of 20 years or more; and climate change initiatives easily may span several generations, yet little attention typically is given to understanding how people's values and perceptions (i.e., their anticipated as compared with experienced utility) may change over this same period. Further work on these topics is urgently needed to understand appropriate societal responses to some of the most serious (and more controversial) environmental policy debates such as global climate change or species vulnerability.

Even when time is considered explicitly, such as with discounting of streams of costs and benefits to determine a present economic value, the typical practice is to employ a single, constant rate of time preference. Yet, as shown by recent descriptive studies, serious questions exist about the applicability of a single discount rate to near term and more distant times (Benzion, Rapaport, and Yagil, 1989), to multiple types of effects (financial, environmental, health) (Chapman and Elstein, 1995), and to benefits (i.e., gains) as well as costs (i.e., losses) (Loewenstein and Prelec, 1992). Other factors, such as the embedding of outcomes in sequences (Loewenstein and Prelec, 1993) and accounting for the uncertainty associated with either the anticipated effects or the changes in future discount rates (Newell and Pizer, 2003), also can influence the choice of appropriate intertemporal environmental policies. Although questions relating to the evaluation of time affect many environmental initiatives, the research conducted to date rarely has emphasized either prescriptive or normative implications; one result is that its influence on the practice or thinking of resource managers has been limited.

CONCLUSION

These three broad areas within environmental decision making all contain high-priority topics for research. In some cases, the further investigation of ideas and techniques would be quite straightforward; research already is well under way, and progress would involve application of known ideas to new areas or the formation of new bridges across disciplinary lines. In other cases there exist major barriers to the conduct of research or to the implementation of new ideas; these include institutional constraints, high levels of uncertainty, or fundamental conflicts among opposing views. The discussion in this appendix highlights these opportunities and barriers.

The potential users of the information that could be produced by a greater emphasis on research along the lines outlined here include all those individuals, small groups, and organizations that are faced with tough environmental choices. We started by emphasizing that decisions at all three levels of social decision making share the difficulty of addressing complexity in environmental decisions in responsible ways. All three of these levels interact, and opportunities and methods of improved decision making at one level can help inform and shape decision making at other levels.

It is hoped that outcomes such as lower environmental compliance costs, improved environmental protection, and increased community acceptance would accompany improved environmental decision making and provide visible evidence of substantive results. Indirect benefits could include better understanding of and preparation for low-probability high-consequence events, less time energy and money spent in the courts contesting regulatory decisions, and an enhanced belief and trust in the wisdom and common sense of decision makers.

Overall, we believe that environmental problems provide an excellent example of how society is attempting to deal with conflicts that involve multiple interests and complex technical content. When environmental decision-making processes succeed, a mechanism is provided for the orderly sharing of views and for the bridging of disciplinary gaps. When they fail, the door is opened to litigation, economic hardship, and the imposition of political solutions that often need to be revisited in a surprisingly short time. It is therefore important that the promise of sound environmental decisions be supported as fully as possible; in this appendix we have tried to outline some of the research and methods that might best achieve that goal.

ACKNOWLEDGMENTS

The authors appreciate the guidance and encouragement of Paul Stern and two anonymous referees in completing this paper.

REFERENCES

Argote, L.
1999 *Organizational Learning: Creating, Retaining, and Transferring Knowledge.* Boston, MA: Kluwer.

Argyris, C., and D. Schon
1978 *Organizational Learning: A Theory of Action Perspective.* Reading, MA: Addison-Wesley.

Arvai, J.L.
2003 Communicating the results of participatory decisionmaking: Effects on the perceived acceptability of risk-policy decisions. *Risk Analysis* 23:281-291.

Baron, J., and M. Spranca
1997 Protected values. *Organizational Behavior and Human Decision Processes* 70:1-16.

Beierle, T., and J. Cayford
2002 *Democracy in Practice: Public Participation in Environmental Decisions.* Washington, DC: Resources for the Future.

Bell, D., H. Raiffa, and A. Tversky
1988 Descriptive, normative, and prescriptive interactions in decision making. Pp. 9-30 in *Decision Making: Descriptive, Normative, and Prescriptive Interactions,* D. Bell, H. Raiffa, and A. Tversky, eds. New York: Cambridge University Press.

Benzion, R., A. Rapaport, and J. Yagil
1989 Discount rates inferred from decisions: An experimental study. *Management Science* 35:270-284.

Canadian Standards Association
1997 *Risk Management Guidelines for Decision Makers.* Ottawa, Ontario: Canadian Standards Association.

Cash, D.W., and S.C. Moser
2000 Linking global and local scales: Designing dynamic assessment and management processes. *Global Environmental Change* 10:109-120.

Chapman, G., and E. Elstein
1995 Temporal discounting of health and money. *Medical Decision Making* 15:373-386.

Chess, C., and K. Purcell
1999 Public participation and the environment: Do we know what works? *Environmental Science & Technology* 33:2685-2691.

Cullen, A., and M. Small
2004 Uncertain risk: The role and limits of quantitative assessment. Pp. 163-212 in *Risk Analysis and Society,* T. McDaniels and M. Small, eds. New York: Cambridge University Press.

Damasio, A.
1994 *Descartes' Error: Emotion, Reason, and the Human Brain.* New York: Avon.

Duram, L., and K. Brown
1999 Assessing public participation in U.S. watershed planning initiatives. *Society and Natural Resources* 12:455-467.

Environmental Protection Agency
2001 *Improved Science-Based Environmental Stakeholder Processes.* (EPA-SAB-EC-COM-01-006.) Washington, DC: Environmental Protection Agency.
Fiske, A., and P. Tetlock
1997 Taboo trade-offs: Reactions to transactions that transgress spheres of justice. *Political Psychology* 18:255-297.
Frewer, L., S. Hunt, M. Brennan, S. Kuznesof, M. Ness, and C. Ritson
2003 The views of scientific experts on how the public conceptualize uncertainty. *Journal of Risk Research* 6:75-85.
Glavin, T.
2000 *The Last Great Sea.* Vancouver, British Columbia: Greystone Books.
Gregory, R.
2002 Incorporating value tradeoffs into community-based environmental risk decisions. *Environmental Values* 11:461-488.
Gregory, R., and L. Failing
2002 Using decision analysis to encourage sound deliberation: Water use planning in British Columbia, Canada. *Journal of Policy Analysis and Management* 21: 492-499.
Gregory, R., B. Fischhoff, and T. McDaniels
2004 Acceptable Input: When Are Deliberative Processes Good Enough? Unpublished manuscript.
Gunderson, L., and C.S. Holling, eds.
2002 *Panarchy.* Washington, DC: Island Press.
Gunderson, L., C.S. Holling, and S. Light, eds.
1995 *Barriers and Bridges to the Renewal of Ecosystems and Institutions.* New York: Columbia University Press.
Hammond, J., R.L. Keeney, and H. Raiffa
1999 *Smart Choices: A Practical Guide to Making Better Decisions.* Cambridge, MA: Harvard Business School Press.
Holling, C.S., ed.
1978 *Adaptive Environmental Assessment and Management.* New York: John Wiley & Sons.
Jasanoff, S.S.
2002 Citizens at risk: Cultures of modernity in the U.S. and E.U. *Science as Culture* 11:363-380.
Jennings, P.D., P.A. Zandbergen, and M. Martins
2001 Complications in compliance: Variations in enforcement in British Columbia's Lower Fraser Basin, 1985-1996. In *Organizations, Policy, and the Natural Environment: Institutional and Strategic Perspectives,* A. Hoffman and M. Vantresca, eds. Palo Alto, CA: Stanford University Press.
Kahneman, D., P. Slovic, and A. Tversky, eds.
1982 *Judgment Under Uncertainty: Heuristics and Biases.* New York: Cambridge University Press.
Keeney, R., and T. McDaniels
2002 A framework to guide thinking and analysis about climate change policies. *Risk Analysis* 21(6):989-1000.
Keeney, R., and H. Raiffa
1993 *Decisions with Multiple Objectives.* New York: Cambridge University Press.
Keeney, R., and D. von Winterfeldt
1991 Eliciting probabilities from experts in complex technical problems. *IEEE Transactions on Engineering Management* 33:83-86.

Keeney, R.L.
1992 *Value-Focused Thinking: A Path to Creative Decision Making*. Cambridge, MA: Harvard University Press.
Kleindorfer, P., H. Kunreuther, and P. Schoemaker
1993 *Decision Sciences*. New York: Cambridge University Press.
Lee, K.N.
1993 *Compass and Gyroscope: Integrating Science and Politics for the Environment*. Washington, DC: Island Press.
Levitt. B., and J. March
1988 Organizational learning. *Annual Review of Sociology* 14:319-340.
Loewenstein, G.
1996 Out of control: Visceral influences on behavior. *Organizational Behavior and Human Decision Processes* 65:272-292.
Loewenstein, G., and D. Prelec
1992 Anomalies in intertemporal choice: Evidence and an interpretation. *Quarterly Journal of Economics* 107:573-597.
1993 Preferences for sequences of outcomes. *Psychology Review* 100:91-108.
MacGregor, D., and P. Slovic
1986 Perceived acceptability of risk analysis as a decision-making approach. *Risk Analysis* 6:245-256.
March, J.
1997 *A Primer on Decision Making: How Decisions Happen*. New York: Basic Books.
McDaniels, T., and R. Gregory
2004 Learning as an objective within a structured risk management decision process. *Environmental Science & Technology* 38(7):1921-1926.
McDaniels, T., and K. Thomas
1999 Eliciting public preferences for local land use alternatives: A structured value referendum with approval voting. *Journal of Policy Analysis and Management* 18(2):264-280.
Mellers, B., V.M. Richards, and M. Birnbaum
1992 Distributional theories of impression formation. *Organizational Behavior and Human Decision Processes* 51:313-343.
Mellers, B., A. Schwartz, and A. Cooke
1998 Judgment and decision making. *Annual Review of Psychology* 49:447-477.
Morgan, G., and M. Henrion
1990 *Uncertainty*. New York: Cambridge University Press.
National Research Council
1996 *Understanding Risk: Informing Decisions in a Democratic Society*. P.C. Stern and H.V. Fineberg, eds. Committee on Risk Characterization. Commission on Behavioral and Social Sciences and Education. Washington, DC: National Academy Press.
Newell, R., and W. Pizer
2003 Discounting the distant future: How much do uncertain rates increase valuations? *Journal of Environmental Economics and Management* 46:52-71.
Norton, B., and A. Steinemann
2001 Environmental values and adaptive management. *Environmental Values* 10:473-506.
Parson, E., and W. Clark
1995 Sustainable development as social learning: Theoretical perspectives and practical challenges for the design of a research program. In *Barriers and Bridges to the Renewal of Ecosystems and Institutions*, L. Gunderson, C.S. Holling, and S. Light, eds. New York: Columbia University Press.

Payne, J., J. Bettman, and E. Johnson
1992 Behavioral decision research: A constructive processing perspective. *Annual Review of Psychology* 43:87-131.
1993 *The Adaptive Decision Maker*. New York: Cambridge University Press.
Pidgeon, N., and R. Gregory
2004 Judgment, decision making, and public policy. Pp. 604-623 in *Handbook of Judgment and Decision Making*, D. Koehler and N. Harvey, eds. Oxford, England: Blackwell.
Presidential-Congressional Commission on Risk
1997 *Risk Management*. Washington, DC: The White House.
Renn, O., T. Webler, and P. Wiedemann
1995 *Fairness and Competence in Citizen Participation*. Dordrecht, The Netherlands: Kluwer.
Royal Commission on Environmental Protection
1998 *Setting Environmental Standards*. London, England: Royal Commission on Environmental Protection.
Saaty, T.L.
1991 *Multicriteria Decision Making: The Analytical Hierarchy Process*, extended ed. Berlin: RWS Publishers.
Satterfield, T., P. Slovic, and R. Gregory
2000 Narrative valuation in a policy judgment context. *Ecological Economics* 34: 315-331.
Slovic, P.
1995 The construction of preference. *American Psychologist* 50:364-371.
Slovic, P., B. Fischhoff, and S. Lichtenstein
1976 Cognitive processes and societal risk taking. In *Cognition and Social Behavior*, J.S. Carroll and J.W. Payne, eds. Potomac, MD: Erlbaum.
Slovic, P., M. Finucane, E. Peters, and D. MacGregor
2002 The affect heuristic. In *Heuristics and Biases: The Psychology of Intuitive Judgment*, T. Gilovich, D. Griffin, and D. Kahneman, eds. New York: Cambridge University Press.
Tversky, A., and D. Kahneman
1981 The framing of decisions and the psychology of choice. *Science* 211:453-458.
von Winterfeldt, D., and W. Edwards
1986 *Decision Analysis and Behavioral Research*. Cambridge, England: Cambridge University Press.
Walters, C.J.
1986 *Adaptive Management of Renewable Resources*. Caldwell, NJ: Blackburn Press.
Weimer, D., and A. Vining
1999 *Policy Analysis: Concepts and Practice*, Third Edition. Englewood Cliffs, NJ: Prentice Hall.

Appendix C

Business Decisions and the Environment: Significance, Challenges, and Momentum of an Emerging Research Field

Andrew J. Hoffman

INTRODUCTION

Over the past four decades, the concept of corporate environmentalism was born and redefined through multiple iterations. Driven by major environmental events such Rachel Carson's *Silent Spring*, the Santa Barbara oil spill, the Cuyahoga River fire, Love Canal, Bhopal, the *Exxon Valdez* spill, the Brent Spar controversy, and many others, conceptions of corporate environmentalism as regulatory compliance in the 1970s gave way to newer management conceptions of "pollution prevention," "total quality environmental management," "industrial ecology," "life-cycle analysis," "sustainable development," "environmental strategy," "environmental justice," and others. The media focus of these conceptions expanded from air and water in the 1970s to hazardous waste, remediation, toxics, right to know, ozone, global warming, acid rain, solid waste, chlorine phase out, and environmental racism today. And with each conception came greater complexity for understanding the intersection of business activity and environmental protection. In particular, empirical data since the 1980s has raised questions about whether environmental protection and economic competitiveness can, at times, be complimentary (i.e., Sarokin, Muir, Miller, and Sperber, 1985).

Concurrent with this evolution in corporate practice has been the emergence of academic research focused on business decision making, firm behavior, and the protection of the natural environment. Among the academic sciences this is a relatively new field, coming into being only in the early 1980s with articles addressing the overlap between business strategy and the environment (i.e., Royston, 1979, 1980) and, later, with the formation of research consortia such as the Greening of Industry Network (Fischer and Schott, 1993) and the Management Institute for Environment and Business (now part of the World Resources Institute). What began as a modest offshoot of management research has grown into a maturing area of study within the management sciences. It is now possible to step back and view the state of this field in terms of where it has been and where it is going. In this appendix I consider what is distinct about existing research in business decision making and the environment and consider future directions in which the field is going.

CORPORATE ENVIRONMENTALISM AS AN EMPIRICAL DOMAIN

The past century has witnessed unprecedented economic growth and human prosperity. Global per capita income has nearly tripled (World Business Council on Sustainable Development, 1997), average life expectancy has increased by almost two-thirds (World Resources Institute, 1994), and people are significantly more literate and educated than their predecessors. Many of these improvements in the quality of life have been driven by the accomplishments of industry. Advancing developments in medicine, materials, transportation, communication, and food production have all emerged from the industrial sector. But, since the 1960s, society has begun to question some of the assumptions around the treatment of the environment as (1) an endless source of resources and (2) a limitless sink for wastes. This has resulted in both an appreciation that corporate activity is the source of environmental problems, but also more recently that industry can also be the solution. This is the area in which research in managerial decision making has the most to offer.

Industry as a Problem

In 2000 private worldwide consumption expenditures reached more than $20 trillion, an increase of more than fourfold since 1960 (in 1995 dollars) (Starke, 2004). To fuel this consumption, industry consumes vast amounts of material resources, and rates will increase. Between 1990 and 2000, sales of the largest 100 transnational corporations increased 50 percent to $4.8 trillion (World Resources Institute, 2001). And 50-75 percent of the annual resource inputs to industrial economies overall become wastes within a year (World Resources Institute, 2000a). This industrial activity has had and will continue to have critical impacts on the natural environment. For example:

- The global rate of deforestation averaged 9 million hectares per year in the 1990s (World Resources Institute, 2001). Global wood consumption has risen by 64 percent since 1961. During that time, half of that consumption was burned as fuel, and commercial logging has cleared more than one-fifth of the world's entire tropical forest cover. Demand for industrial wood fiber is projected to rise between 20 and 40 percent by 2010 (World Resources Instutute, 1999).

- Soil degradation has become a major issue on as much as 65 percent of agricultural land worldwide, reducing the productivity of about 16 percent of that cropland, especially in Africa and Central America (World Resources Institute, 2000b).
- Consumption of fish and fishery products (such as fish meal and fish oils) has risen by 240 percent between 1960 and 2003 and more than fivefold since 1950 (World Resources Institute, 1999). In 1999 the global total fish catch was 4.8 times higher than in 1950. Since that time, industrial fleets have exhausted at least 90 percent of all large ocean predators (such as tuna, marlin, and swordfish) (Starke, 2004). Overall, nearly 25 percent of the world's most important marine fish stocks have been depleted, overharvested, or are just beginning to recover from overharvesting. Another 44 percent are being fished at their biological limit and are, therefore, vulnerable to depletion (World Resources Institute, 2000b). Demand for food fish is projected to increase by 34-50 percent by 2010, a level of consumption that cannot be met if current production trends continue unchanged (World Resources Institute, 1999).
- Global use of fossil fuels has increased 4.7 times between 1950 and 2002 (Starke, 2004), such that worldwide emissions of the greenhouse gas, carbon dioxide (CO_2), have increased to 23 billion metric tons in 1999, an 8.9 percent increase since 1990 (World Resources Institute, 2003). This rise is expected to yield changes in global weather patterns, increases in sea levels, and the migration of vectorborne diseases.

Industry as a Solution

The above examples illustrate how business activities impact the environment in significant ways. And as these types of environmental degradation continue to grow, companies will experience more and more pressure to find solutions. Although in truth it is industrial society as a whole that causes environmental problems, it is industry that will bear the burden of reducing their severity. Empirically, as the world becomes more globalized, and the impact of industrial and commercial activities become more extreme, no solution to the environmental problems society faces will be solved without the involvement of business. The reasoning for this assertion follows several lines.

First, business decisions about what inputs to use and how to manage outputs ultimately determine environmental quality. Therefore, industry often bears the direct cause-and-effect link to environmental problems, such that it is the most vulnerable institution to social and political challenges for change. Second, firms are, in general, the sources of technological evolution within society. As such, in many cases firms best understand the technical trade-offs that innovation choices may involve. Although environmentalists and others may appreciate impacts of systemic change, firms understand the underlying technical and economic aspects of innovative activities. Third, governments no longer possess the full array of resources and knowledge necessary to dictate environmental solutions to business. Many within policy circles now agree that business must become a participant in the environmental regulatory process if sustainable and economically efficient solutions are to be found. And fourth, the power of business organizations to determine the structures of our social, economic, and political activity has grown to such enormous proportions that industry now possesses the most resources both individually and through markets to create more efficient coordinating mechanisms. And indeed, business has been developing solutions to emerging environmental problems with products and services such as alternative mobility systems such as gas-electric hybrid vehicles, fuel cell vehicles, or car sharing in urban centers; alternative energy sources such as wind energy, fuel cells, or microturbines; alternative materials such as biomaterials (to replace fossil-fuel-based fabrics such as nylon, polyester, and Lycra) or composite woods to replace large stock timber.

Such solutions can best be found when the industrial sector works in concert with other sectors of society (Dietz, Ostrom, and Stern, 2003). As a result, there is a great need for the study of business decision making as part of a social science research agenda. At the core of this agenda are some simple and straightforward questions: What is the relationship between environmental protection and corporate competitiveness? How does this relationship alter the basic elements of corporate management and objectives? How can we anticipate future ideas of the objectives, purposes, and practices of the corporate organization in light of emerging concerns for environmental protection? Scholars within business schools are now striving to understand the implications of these questions. And importantly, they bring a distinct set of capabilities, models, and theories toward answering them.

Business Challenges

As an empirical domain, corporate environmentalism comprises a blend of characteristics that make it distinct from other pressures with which the firm is familiar, necessitating a distinct research domain. It has many characteristics similar to other social issues such as gender equity, affirmative action, or labor relations, although it is also distinct from these issues in several ways. On the other hand, it has technical and economic components that make it similar to other strategic issues such as consumer demand, material processing, or competitive strategy, but again it has differences that require special attention. For the corporate organization and the manager, it is the issue's ability to *merge the social and the technical* in its impact on corporate practice that makes environmentalism unique.

The Social Dimensions of Corporate Environmentalism

On its most fundamental level, environmentalism is a social movement much like gender equity, civil rights, and labor relations. It has constituent groups that lobby for social change on all levels of society. However, the makeup of this constituency is more troublesome for the corporation than that of other social movements. Membership in the environmental movement is indeterminate. In settling issues of labor relations, managers negotiate with workers and union officials. In settling issues of civil rights or gender equity, there are female, minority workers, and national organizations set up to represent them. However, with the environment there are few natural constituency or bearers (Buttel, 1992). A high-quality environment tends to be a public good, which when achieved cannot be denied to others, even to those who resist environmental reforms. For many environmental issues, those who act to protect the environment can expect to receive no personal material benefits (Buttel, 1992). So the firm is left to decide who is a legitimate representative for environmental concerns.

Often those representatives are organized environmental nonprofit groups. But the indeterminism of environmentalism also means that it attracts a wide range of supporters cutting across social, economic, and demographic lines. Those representing environmental interests to the firm or to society at large go beyond the nongovernmental organization community. Others, such as employee groups, labor unions, community groups, consumers, environmental activists, investors, insurers, the government, and industry competitors have become active environmental advocates. Even internal managers can become advocates for the environment (Morrison, 1991:18). Interacting with such a wide range of interests has necessitated new structures and internal conceptions of the firm's organization and purpose.

Furthermore, the social issue of environmentalism has a decidedly nonsocial component. More than just a constituency of social advocates, there is also the environment itself to contend with. The prominence and power of environmental change (and in the most extreme case, environmental catastrophes) act as another form of social pressure, placing demands on our social, political, economic, and technical structures that are unique from any other demands the corporation faces. They focus attention without warning, imposing demands for action and change. While open to social interpretation and enactment (Hoffman and Ocasio, 2001), environmental events force corporations, government, and activists to devote resources and attention to the environmental issue.

The Technical Dimensions of Corporate Environmentalism

Where issues such as affirmative action and gender equity transcend industries and have little direct affect on production processes or product development, environmentalism has a distinct technical component, directly challenging how corporations handle material resources and produce goods and services. Over the past three decades, the technological demands for corporate environmental responsibility have shifted from removing only visible levels of contaminants from effluent streams to now removing concentrations in the parts per billion range and, at times, parts per trillion. Beyond process emissions, environmentalism also mandates changes in the content of product development. New laws mandating the public disclosure of emission levels and product contents as well as the potential health effects of those chemicals create daunting technological challenges for the firm (Hoffman and Ehrenfeld, 1998). The effects of these demands are not universal. Some industries, such as oil and chemicals, face greater challenges in both the measurement and the control of hazardous emissions. Even within industries, different companies face differential challenges in developing new products, processes, or raw materials in the face of environmental demands. The technical challenges of environmentalism add a new dimension to the strategic landscape, one that will often decide which firms will succeed and which will fail (Hannan and Freeman, 1977).

Often, firms are required to collect data, initiate change, and develop an understanding of their processes and products at levels that are not considered necessary for traditionally accepted strategic reasons. This is because strategy and technology are socially influenced by constituents outside the firm. Engineers can no longer focus simply on the end-based results of engineering calculations. They must now understand the social, political, economic, and cultural context of their task. Environmentalism signifies a redefinition of both technology and the corporation's role in developing it. New concepts such as waste minimization, pollution prevention, and product stewardship are finding their way into all aspects of operations, from process design to product development.

Beyond conceptions of technology, environmentalism challenges economic conceptions of the firm. Unlike other social issues that deal with equity and the fair distribution of opportunity and wealth, environmentalism increasingly affects basic business economics, effectively redefining the conceptions of production in industry. Issues such as gender equity or affirmative action will involve some gain or loss to specific individuals within the firm; however, the economic output of industrial activity should remain fundamentally unchanged (Hoffman and Ehrenfeld, 1998). Social issues bear more commonly on issues of sharing what we have, issues of social equity.

But environmentalism produces a different outcome. Environmentalism interferes with fundamental economic models of consumption and production, resulting in a net change in efficiency. For example, a recent debate has emerged over the economic impact of climate change controls. Some estimates predict a drain on the gross national product by as much as 3.5 percent if aggressive emission reduction targets are set. Others estimate that modest controls on greenhouse gas emissions would not damage the economy, that the world has significant opportunities to control emissions by making its energy systems and automobiles more efficient. This more efficient use of energy is estimated to increase the gross national product by 1 or 2 percent. Such a debate would not accompany new laws regarding racial or gender equity.

For the individual firm, the impact is no less direct. Environmental concerns can cause the elimination of entire product markets, such as those for chlorofluorocarbons, DDT, and dioxin. They can also cause the formation of new markets as they did for Freon substitutes, termed hydrochlorofluorocarbons, in the wake of the 1987 worldwide ban on chlorofluorocarbon production. Finally, environmental liability has risen to levels that have shaken the basic precepts of corporate risk management. Most notably, the $5 billion in fines and penalties against the Exxon Corporation for the 1989 *Valdez* spill would have bankrupted many other "smaller" corporations. Regardless, the threat of such large fines has caused most firms to alter their oil transport strategies.

In essence, what has evolved is an alteration of the core objectives of the firm and the basic conceptions of production. Shareholder equity may remain the single most important criteria for corporate survival. Yet environmental responsibilities are infiltrating the taken-for-granted beliefs that have previously guided that pursuit. Today, most U.S. companies have a formal system in place for proactively identifying key environmental issues as part of their overall corporate strategy (Morrison, 1991). Environmental strategy incorporates a merger of these social considerations with the technical aspects of corporate operations.

CORPORATE ENVIRONMENTALISM AS A RESEARCH DOMAIN

The study of business and the natural environment lies at a unique juncture of the physical and the social sciences, scientific disciplines that seek to understand the behavior of natural ecosystems either as separate entities or in their relation to social systems. The way we understand these systems as separate entities is through the physical sciences of chemistry, toxicology, biology, physics, entomology, and others. In fact, the study of the environment has been on the agenda of the modern physical sciences for long enough that boundary-spanning research specialties like ecology are now recognized areas of research and professional standing.

In contrast, attention to the natural environment within the social sciences is relatively preliminary both in research traditions and in professional infrastructure and has few established cross-disciplinary research fields (efforts in this regard are noted in the areas of urban planning, geography, and risk management). Subspecialties in many social science disciplines and associated professional fields such as law, economics, philosophy, theology, ethics, sociology, psychology, and political science do focus on environmentalism, each investigating the linkages between social and environmental systems in its own specialty idiom of characteristic research questions, designs and evidence, and implications. Each of these offers a different vantage point, allowing for a contribution to a complementary synthesis of ideas for explaining social and organizational behavior (Allison, 1971) as it relates to the natural environment. Below is a review of five disciplinary vantage points followed by a discussion in greater depth on how the environment is viewed from the field of business management. This review is not meant to be an exhaustive assessment of the breadth and depth of each discipline. Rather it is meant to highlight some of the influential and potentially productive areas of study as it relates to business activity. It is also an attempt to show the variety of research undertaken in these disciplines as background for the more specific research being conducted within the more focused management disciplines.

Perspectives from Economics

Scholars within the field of economics cover a variety of topics including the valuation of natural ecosystems and resources, analyses of social costs and benefits, the creation of market mechanisms to alter polluting activities, bounded rationality, the economics of innovation, agglomerated economies, and organizational behavior. Those addressing issues of corporate decision making tend to consider the nature of pollution and the environment with a long-standing set of policy approaches focusing on "market failures" (such as "externalities" and imperfect or asymmetric information about risks) and "public goods." In this domain, environmental damages that are imposed on downstream or downwind residents or the public at large are often omitted from market prices and thus treated as "free" to the producers and consumers that cause them. Public goods—even essential environmental services for which no markets exist, such as clean air and other "common pool resources"—also are often destroyed because excessive or damaging uses cannot easily be excluded, and each user tends to undervalue them. And many natural assets such as petroleum and ancient forest stocks are priced only at their value in current markets, omitting their potentially greater value as sustainable capital assets. The harm caused by these outcomes is the "consequence of an absence of prices for certain scarce environmental resources (such as clean air and water)" (Cropper and Oates, 1992:675). Left unregulated, economists observe that private firms do not choose "socially efficient" levels of environmental protection (Tietenberg, 1992). They "externalize" these environmental costs and thus avoid paying the full social costs of the environmental damage they cause (Baumol and Blinder, 1985). To provide the needed signals for correcting the market and providing economic incentives for good environmental behavior, economists prescribe the introduction of surrogate prices such as unit taxes, effluent fees, or tradable credits (Hahn and Stavins, 1991).

Perspectives from Ethics

Scholars in the field of ethics focus on the nature and morality of human conduct. When addressing corporate activity as it relates to the environment, this field focuses on the role of the corporation within society and its responsibilities toward conserving, preserving, and utilizing natural resources. It mixes descriptions of what presently *is* with prescriptions of what *ought* to be. It is a normative discovery of human values derived from science, metaphysics, aesthetics, epistemology, philosophy, and judgments of intrinsic values (Hargrove, 1989). Where these fields have traditionally concerned themselves with an account of the goods of culture and of the right and wrong of interpersonal relations between man and man, environmental ethics takes traditional ethics one step forward, acknowledging that humans inhabit natural communities and expanding ethics to consider human responsibility for nature (Holmes, 1988). More specifically, it argues the thesis that human populations, nonhuman animals, and nonsentient nature are all morally considerable. They may not be counted by the same metric, but each counts in moral calculations because each has intrinsic value. Where traditional ethics places man at the center of the moral universe, environmental ethics expands the scope of that universe and man's place within it (Eliot and Gore, 1983). Of particular importance in this discussion is the place of the corporation—man's dominant instrument for utilizing natural resources—within the natural environment.

Perspectives from Law

When addressing concerns over corporate activity and the natural environment, scholars in the field of law focus on the equitable distribution of rights and liabilities. The legal system is devoted to avoiding or rectifying perceived wrongs that are the result of human or nonhuman action. It is the product of a society's collective and conflicting values, which are incorporated with scientific knowledge and are reflected in laws. The (American) legal system is built on the foundation of common-law decisions and principles, which is overlain with a later statutory system that attempts to correct the deficiencies of the earlier one. Decisions are the product of logical arguments based on legal precedent and supporting evidence. The focus of these decisions is on the property and personal rights of citizens. These rights include the rights to use the property we own in the manner that we chose; the right to enjoy our own property without unreasonable trouble from our neighbors; and finally, the right we have (or think we have) to a "decent environment" in which to live (Hoban and Brooks, 1996; Revesz, 1997). Over time, longstanding common-law precedents protecting individuals from upstream or upwind polluters were supplanted by judicial doctrines of "reasonable use" that favored industrial polluters; and as environmental damage subsequently increased, new environmental regulatory statutes provided limited substitutes for portions of these early precedents, but often in forms that prescribed costly and rigid (although easily enforceable) end-of-pipe technological controls rather than more efficient performance-based incentives.

Perspectives from Business History

Historical studies search for explanatory power in events, actions, and stories. Traditional business history studies focus on organizational, cultural, and strategic considerations within organizational decision-making processes, largely defined by the work of Chandler (John, 1997; Galambos, 1970). This work has provided evidence of variable concerns among managers and firms, publics and special constituencies, and governmental actors that dates from at least the mid-nineteenth century (Rosen, 1995, 1997). Standard emphasis among environmental historians has dealt with wilderness, the conservation movement, or the modern environmental movement. But more recently, these fields have begun to merge, identifying concerns at the intersection of business, markets, and environmental change (Cronon, 1991; Hays, 1998; McGurty, 1997; Andrews, 1999). "[F]rustrated with environmental history's longstanding focus on farms, forests, and wilderness and fortified by a dawning recognition of the much wider scope of the 'natural,' many environmental historians have begun to gravitate away from the study of pristine environments toward those more thoroughly and unmistakably shaped by human hands" (Rosen and Sellers, 1999:582-583). This refocus on the environmental dimensions of industrial development is evident in recent studies, symposia, and review essays that chart new questions and new collaborations between business historians and environmental historians. This emerging tradition of research focuses on "physical processes by which the stuff of nature—'raw' materials—was carved or coaxed out of mountains, forests, and deserts, channeled into factories and squeezed and cajoled into commodities . . . varieties of 'waste' generated by business and customers; . . . to the effects of resource extraction and use" (Rosen and Sellers, 1999:577). This approach weaves business together with its material and symbolic environments in a seamless web, a basis to bring complex physical, cultural, managerial, technological, and economic connections between business and the environment into better focus and hence to explore business in relation to public policy.

Perspectives from Sociology

Organizational and sociological study of the interaction between the natural environment and social organization and behavior dates at least from the early 1970s, coinciding with the emergence of environmental activism and social movements in the United States, Europe, and elsewhere (Laclau and Mouffe, 1985). This is evident in activity in professional associations, intellectual organizing, and specialty journals. By the mid-1970s, the American Sociological Association, the Rural Sociological Association, and the Society for the Study of Social Problems had all established sections related to environmental sociology (Dunlap and Catton, 1979). To provide an outlet for this growing volume of research, special journal issues were devoted to environmental sociology: *Sociological Inquiry* (1983), *Annual Review of Sociology* (1979, 1987), *Journal of Social Issues* (1992), *Qualitative Sociology* (1993), *Social Problems* (1993), *Canadian Review of Sociology and Anthropology* (1994) (Hannigan, 1995). Schools increasingly posted position announcements in environmental sociology, and numerous research centers and institutes were established, including targeted funding for dissertations and some postdoctoral funding such as the National Science Foundation program initiatives in the early 1990s on global environmental change.

By the late 1980s, reviews of the field identified five areas of scholarship in environmental sociology (Buttel, 1987): (1) new ecological paradigm; (2) environmental attitudes, values, and behaviors; (3) the environmental movement; (4) technological risk and risk assessment; and (5) the political economy of the environment and environmental politics. By the mid-1990s, a focusing of the research agenda included several streams of importance to business decision making. The new ecological paradigm (Catton and Dunlap, 1980)—the shift away from anthropocentric (human-centered) to ecocentric thinking (humans are one of many species inhabiting the earth)—has become an influential theoretical insight of environmental sociology, one that has been picked up by several management-oriented scholars such as Gladwin, Kennelly, and Krause (1995). Other researchers deal with concerns for the political and economic root causes of environmental disruption and the development of a systematic approach that shows how organizations, institutions, and individuals can push for environmental protection reforms (Schnaiberg and Gould, 1994) and competing conceptions of nature and analyses of how those conceptions have emerged (Cronon, 2003; Botkin, 2004). And still others attend to the rise of environmental consciousness and social movements (McAdam et al., 1996), addressing how change occurs within social systems and why. Central to this stream is a consideration of environmental risks as they relate to the macrosociology of social change (Beck, 1992). The field now appears to be centering on a social constructionist approach to addressing these key themes that focuses on the "social, political, and cultural processes" by which environmental issues, problems, and solutions are given attention and defined (Hannigan, 1995:30). This remaking of the focus of the subfield raises a perennial tension between the intellectual goal to foster research in the subfield and the professional project of defining a distinct stand-alone empirical field for research. At the root of this tension is the added value of creating distinct specialty fields versus remaining engaged with wider disciplinary approaches.

In sum, each of these disciplinary perspectives describes quite distinct characteristic concerns. In each, the study of environmentalism is described in the standard terms of the discipline. In each, there are scholars working at the edge of the discipline in order to take advantage of the distinct features of environmentalism as a theoretical and empirical pivot for further research. Each intellectual tradition approaches the issue from a different angle, using different terminology, asking different questions, and yielding different answers. Each also has a set of voices making links between the disciplinary standards, research, and policy and practice issues. The concerns and research infrastructure of environmentalism in organizations, strategy, and management look different.

CORPORATE ENVIRONMENTALISM AS A MANAGEMENT DISCIPLINE

Initiatives to Build a Research Community

Scholars in management schools have more recently entered this research domain as well. An international interest group of scholars, the Greening of Industry Network, was formed in 1989. This group produced one of the first collections of research in environmental management. Greening of Industry Network participants argued that "most regulation has not been based on a solid understanding of how industrial firms operated" and that future advances in environmental policy required an appreciation for the "intradynamic and interdynamic processes" of organizational learning that incorporate an awareness for how "various groups both inside and outside the firm conjointly shape its behavior and strategy" (Fischer and Schott, 1993:372).

This first initiative to build a research community among management scholars was followed by the formation of the Management Institute for Environment and Business in 1990 (now a division of the World Resources Institute) and establishment of the Organizations and the Natural Environment special interest group of the Academy of Management in 1994. To support this burgeoning research area, special issues on the natural environment and organizations have appeared in the *Academy of Management Review* (1995), *American Behavioral Scientist* (1999), and *Academy of Management Journal* (2000). Furthermore, academic journals dedicated to the interface between managerial action and environmental protection also emerged, including *Society and Natural Resources, Business Strategy and the Environment, Social Science Quarterly,* and *Organization & Environment.*

The corpus of research parallels developments in environmental sociology. For example, one common theme has been the shift from an anthropocentric to ecocentric perspective similar to the new ecological paradigm (Colby, 1991; Gladwin et al., 1995; Purser, Park, and Montuori, 1995). But a distinction in this research domain is its primary focus on the behavior of the firm, management research, and management education as a self-evident and unquestioned need. Furthermore, although addressing the fundamental question of why firms respond to ecological issues (Hart, 1995; Lawrence and Morell, 1995; Lober, 1996), much of this research has been normative in focus, focusing on understanding and predicting why and how corporations "can take steps forward toward [being] environmentally more sustainable" (Starik and Marcus, 2000:542). Some researchers have focused on the implications of the shift to an ecocentric perspective for organizations (and corporations in particular) (Starik and Rands, 1995;

Shrivastava, 1995). Others have considered how to merge existing concerns for economic competitiveness with environmental demands to gain market advantage (Schmidheiny, 1992; Smart, 1992; Porter and van der Linde, 1995; Stead and Stead, 1995; Roome, 1998; Sexton, Marcus, Easter, and Burkhardt, 1999). But an underlying tension in this domain parallels that within environmental sociology—the question of whether the goal of this group of researchers is to create a distinct specialty field of management inquiry. Some have argued that academic research in the "organizations and natural environment area" is based on a vision of practice and policy based on new values, attitudes, and behaviors (Starik and Marcus, 2000). Others consider this area to be an empirical domain into which existing theory can be applied. These are fruitful tensions about intellectual and professional strategies. Regardless of this debate, the field is in development and embarking on streams of research in multiple directions.

Emerging Directions in Environmental Management Research

Research within the management sciences on environmental issues falls generally into seven basic areas within the business school community, each with its set of concerns and research tracks.

Strategy

Some of the early research on environmental strategy attempted to show the link between positive environmental performance and positive competitive performance. Questions over whether it "pays to be green" emerged in a cadre of papers (King and Lenox, 2001). Yet a more recent examination has begun to ask, not if, but how and when firms can create competitive advantage through environmental protection (Howard-Grenville and Hoffman, 2003). This is an area of great empirical and theoretical importance and has tremendous linkages to work in the field of entrepreneurship. Toward this end, some research is being performed on the relationship among uncertainty, general environmental factors, resources, and proactive environmental strategies (Aragón-Correa and Sharma, 2003). Significant research demonstrates the relationship between resources and environmental strategies (Shrivastava, 1995; Hart, 1995, 1998). However, little is known about what impacts that relationship. How do organizational and managerial variables as well as stakeholder relationships impact that relationship? Similarly, little research exists on the measurement of critical resources as they impact environmental strategies. The resource-based view of strategy (Hart, 1995; Barney, 1991; Wernerfelt, 1989) is arguably one of the newest and fastest growing areas of strategic inquiry. The resource-based view argues that only resources that are rare, valuable, inimitable, and nonsubstitutable will lead to a sustained competitive advantage. Yet, no adequate operationalization of these resources exists in the context of the environment (or in the minds of some, in all of strategy research).

Another research stream examines what factors—public policy, market and institutional forces, and others—would favor or retard environmentally beneficial innovation, in products as well as production processes, both within and across firms. Toward this end, significant research attention needs to be focused on interfirm collaboration and partnerships toward environmental protection. Oftentimes, the environmental impacts of corporate behavior come from networks of firms operating within a continuous value chain that brings raw materials to final consumption (and sometimes back again). The knowledge and technical expertise in this network does not lie within one single organization but within a constellation of actors that must work together to find solutions (Roome, 1998).

In coordinating this network activity, some have begun to study why firms adopt voluntary standards for environmental performance (Delmas and Terlaak, 2001; Andrews et al., 2001; Delmas, 2002). Others focus on the role of organizational clusters or fields (Jennings and Zandbergen, 1995; King and Lenox, 2000; Bansal and Roth, 2000), interorganizational relationships (Starik and Rands, 1995; Clarke and Roome, 1999), interorganizational alliances (London and Rondinelli, 2003; Rondinelli and London, 2003), and stakeholder relations (Berman, Wicks, Kotha, and Jones, 1999; Clarkson, 1995) as determinants of systemic corporate environmental behavior. Still others prefer to look more carefully at questions about how and why these networks form and what are the coordination mechanisms within them. And finally, these arenas of study can all be addressed at the international level as globalization continues to develop and broaden the impact and possible opportunity for business.

Operations Management

On the level of the individual firm, there is a great need for further research into dematerialization of production processes (Roome, 1998). This can involve optimization of the supply chain logistics for producing goods, developing more efficient manufacturing processes (or related objectives of factor four improvements—Weizacker, Lovins, and Lovins 1998), and utilizing green materials and processes. Or it can involve the shift from products to services in the marketplace (Lovins, Lovins, and Hawken, 1999) such as leased carpets (Interface) or car sharing (Mobility or Zip Car). Continuing this line of inquiry into networks of firms, a great deal of research has been conducted within the domain of industrial ecology since its earliest writings in the late 1980s (Frosch and Gallopoulos, 1989). Using natural ecosystems as its model (Friedman, 2000), industrial ecology highlights transformational change in local, regional, and global material and energy flows, the components of which are products, processes, industrial sectors, and economies. It promotes efficient resource use by reducing environmental burdens throughout the total material cycle. This cycle exists in a continuous feedback loop with materials and energy flowing between natural and industrial systems in three stages: extraction of natural materials, converted into raw materials and mechanical energy; these then worked into usable and saleable products; and finally, these products are distributed, consumed or used, and disposed by consumers. Developed largely by engineers, the central unit of analysis in industrial ecology is that of industrial organizations within broad-scale systems of facilities, regions, industries, and economies and seeks to reduce the environmental burden of that system through broad-scale system-wide changes (U.S. Environmental Protection Agency, 2000). A great deal of research is necessary to understand the linkages among the technical "ecology" of the industrial enterprise and also to incorporate concerns for the "social ecology" into industrial ecology research (Hoffman, 2003).

Organizational Behavior

We now live in an age when environmental concerns originate from a system of pressures much broader than government, activist forces, or supply chains. Increasingly, environmental concerns are becoming infused into the relationships among firms and trade associations, insurance companies, shareholders, investment funds, financial institutions, environmental nongovernmental organizations, the local community, individual citizens, the press, consultants, and employees. Through so complex a systemic web of constituents, environmentalism becomes transformed from something external to the market environment to something that is central to the core objectives of the firm. The definition of what constitutes a "green" company continues to expand as the external pressures for corporate environmental action become more diverse and demanding. More research is necessary for understanding the full dynamics by which this change is taking place; understanding when such change is genuine or a form of greenwashing; analyzing when there are ebbs and flows in this definition with respect to fads and fashions; and covering a range of levels of analysis, including intrafirm dynamics, sectoral dynamics, supply chain dynamics, service platformed on technology dynamics, and global economic systems. For example, research into the ways in which trade associations affect industry-wide change (Nash, 2002) is one avenue. Another is analysis of the ways that overall institutional environments are changing and how this impacts what is expected of firms today and tomorrow on environmental issues (Hoffman, 2001). A related line of inquiry asks how individual firms can influence this change process, in effect playing the role of institutional or social entrepreneur (Lawrence, 1999; Fligstein, 1997).

As corporations respond to this increasing institutional change, they must trigger a more complex set of organizational and strategic responses than merely the management of these external pressures. Scholars approach this issue by analyzing both individual- and organization-level variables. Individual-level variables include concerns like reward systems, selection and socialization, management leadership styles (Egri and Herman, 2000), and individual interpretation and intention (Ramus and Stegner, 2000; Flannery and May, 2000). Organizational variables include concerns such as identity and environmental interpretation (Sharma, 2000), strategic benefits from reputational management (Fombrun and Shanley, 1990), organizational culture (Hunt and Auster, 1990; Roy, 1991), and corporate governance (Kassinis and Vafaes, 2002). Yet much work still needs to be done on understanding how this is done and with what implications for the firm, firm competitiveness, and the motivation of the individual employee. Finally, a great deal of work is necessary for understanding how international culture and corporate greening intersect. As firms become more global in their operations, how do they transfer environmental standards from one national context to another and how do they translate environmental imperatives from one regime to another? U.S. concerns, for example, over endangered species will not resonate with communities in developing countries where their primary concerns may be clean water or proper sewage.

Marketing

When considering the value chain, attention should be paid to the role of the end consumer in driving environmental considerations within the firm. If consumers begin to demand environmental attributes in products, firms will respond to environmental issues as a market opportunity. But pinning down the exact status of environmental consumerism is a difficult challenge. The power of this purchasing block is a much-debated issue. Beyond general attention to the issue, public opinion polls also show that people care about the environment and claim that they will allow that concern to affect their buying decisions (Krupp, 1990; *Times Mirror*, 1995). However, opinion polls and actual buying practices are not tightly linked. It is widely believed that, although they claim otherwise, consumers will not pay a price premium for environmental attributes (Mohr, Eroglu, and Ellen, 1998). Research is necessary to understand the linkages between opinion and behavior.

Research is also necessary for understanding the demographics of green consuming, what drives those consumer-buying decisions, and how to influence or appeal to that decision-making process. Conventional marketing wisdom suggests that the best marketers can expect is that when goods provide comparable value (and are comparably priced), environmental attributes can break the tie. But others are working on designing effective strategies for attracting the consumer to products with green attributes (among others) (Ottman, 1998). For instance, how do people perceive green claims or pressures for behavioral change such as recycling? In product development, how are green issues integrated, "silently" or overtly, into design and development as well as all aspects of marketing planning, especially marketing strategies? And when individual efforts at marketing "green" fail, how can collective efforts be more effective? Marketers are investing in green certification schemes such as the Green Seal, Sustainable Forestry Initiative, or others. More research is necessary for understanding the influence of such schemes on individual buying decisions and the overall value chain. And then the international aspects of green consuming warrant more attention. How do attitudes and willingness to pay on environmental issues differ among consumers in different countries, and what influence does this have for possibilities to integrate environmental aspects into marketing and customer relationships? And all of this leads toward questions regarding greenwashing. Further study is necessary in the understanding of symbolic adoption of green practices or facades in order to gain further market acceptance.

Some suggest that traditional segmentation variables (sociodemographics) and personality indicators are of limited use for characterizing the green consumer (Schlegelmilch, Bohlen, and Diamantopoulos, 1996). This leads some to look beyond demographics to understand how green purchases may be more driven by context and perceived trade-offs (Peattie, 1999). For example, do I drive further to buy the environmentally better products? Do I buy the local organic product, or the fairly traded imports from a poor country? How concerned consumers juggle the different issues in the sustainability agenda and manage the trade-offs between them is an interesting research frontier. Finally, the broadened area of social- (Andreason, 1995) or cause-related marketing (Bloom and Gundlach, 2001) is a vibrant and interesting line of inquiry that deserves further analysis.

Beyond consumers, there is also the area of business-to-business marketing or organizational marketing (as in for government and other public sector purchasing) that gets less attention, but is often where more change is going on in terms of purchase criteria (for example, through the passing of the International Organization for Standardization's ISO 14000 requirements back down a supply chain). So the influence of environmental criteria in industrial and organizational marketing is a key area for research.

Finally, there is research emerging that looks beyond these elements of the green marketing agenda and looks toward more sustainable societies, economies, and companies that will require more significant changes to production and consumption within mass markets rather than market niches. This research agenda seeks to understand how to achieve this end. This research stream considers issues such as problems of marketing to consumers when levels of basic environmental literacy are low; product takeback and the need to engage consumers in the return of old cars and electronics into the supply chain; development of new market structures based around alternatives to purchase and product and service substitutions; design-for-environment products and the use of dematerialization and low-energy products to reduce environmental impacts at no additional cost to the consumer; and the role of marketers as an inhibitor or promoter of environmentally improved products in each of these cases. One critical element is the question of how to communicate means of production issues to consumers effectively.

Accounting

The traditional approach to teaching accounting has been to provide students with a rule-oriented taxonomy where problems fit neatly into specific topical cells. This approach is inadequate for the increasingly complex accounting problems posed by environmental issues (Sefcik, Soderstrom, and Stinson, 1997). Environmental issues challenge accountants to apply existing accounting systems to new settings and to critically analyze existing and proposed accounting systems. Research in this area encompasses economic analysis of incorporation of environmental "externalities" in accounting systems: emerging international standards concerning corporate environmental performance; overhead allocation as strong environmental strategy; reporting rules for environmental liabilities and expenditures; approaches to environmental measurement, cost accounting, and environmental audits; and the impact of information from these systems on decision making (Gentile, 2002). An example is research into full-cost accounting and life-cycle analysis. In essence, how does one incorporate the full environmental costs of a product or process into existing accounting measures and models? Then attention can be applied toward understanding how to link such environmental performance measures to the reward systems within the company in order to motivate environmental behavior.

A different and related line of analysis deals with environmental disclosure. This research has focused particularly on environmental (and now sustainability) annual reports and on the determinants and reasons for more versus less disclosure. The publication of this information is still increasing and taking place increasingly in separate reports, oriented not only at shareholders but also at a range of stakeholders. This raises new questions about the objectives, quality and determinants (country, sector, size, degree of internationalization, multinationality, etc.), and specific drivers (legitimacy, stakeholder management, events) of these reports. Also, it raises challenging questions about the possible liabilities related to disclosing too much or too little information and the extent of accountability and transparency. And finally, new questions about the value and reliability of disclosure in light of recent scandals, (i.e., accusations of "managerial capture") are gaining greater attention (Kolk, 2003; Kolk and van Tulder, 2004).

Finance

Shareholders and investors are powerful forces for change within the corporation. In the cause of the environment, they have been wielding that power since the late 1980s both through shareholder voting and directing capital investment. Beginning in 1989, shareholders began to file environmental proxy resolutions in annual board meetings. However, no one has yet been able to demonstrate conclusively that corporate social responsibility boosts shareholder value, and the evidence is at times conflicting (Margolis and Walsh, 2001). Thus, within the finance community, there is research under way to understand the connections between financial and environmental performance as well as the power and influence of the environmental investor. This power can be a single purchasing block, as through green investment funds, or through the market in general as it reacts to environmental events and issues. "Regular" investors have increasingly raised concerns over the financial risk of environmental issues such as climate change at shareholder meetings, exerting pressure on companies to take measures. In 2002 a shareholder resolution sought to reduce the duties of Lee Raymond, chairman and CEO of Exxon-Mobil, because of his position that climate change was not a problem for the company. The resolution got a surprising 20 percent supporting vote. And this is not the only such resolution. In the 2003 proxy season, there were as many as 19 resolutions filed regarding climate change issues, two-thirds of which received more than 20 percent supporting votes, including GE (22.6 percent), American Standard (29 percent), Eastman Chemical Co. (29 percent) and American Electric Power (27 percent), (Interfaith Center on Corporate Responsibility, 2003). Further study is necessary to understand the trajectory and influence of this activity on corporate actions vis-à-vis the environment. This study should address the assessment of environmental liabilities, the development of risk-return profiles, and then the extent to which a company should disclose such results to the investor community and the public at large. In addition, there has been growth of new market-based solutions to reducing environmental impacts—for example, the U.S. sulfur dioxide emissions permit market (Tietenberg, 2002). A similar market is developing in the European Union. The dynamics and success of these markets provide new areas of interest for finance researchers.

Government Policy

While legal standards have achieved impressive gains in environmental protection and wildlife conservation since the 1960s (Easterbrook, 1995), some argue that the methods they employ are out of date with contemporary environmental problems and that such standards are becoming increasingly inefficient in achieving emerging environmental goals. Existing standards and enforcement programs are perceived to be too rigid and restrictive to foster the type of private innovation (rather than mere compliance) that is required to identify and implement solutions that are both environmentally and economically sustainable (Schmitt, 1994). Believing that we are rapidly approaching the point of diminishing returns on command-and-control environmental regulation, many see the existing policy regime as possibly the greatest obstacle to continued environmental improvement. Some look to the roles of subsidies (often perverse) in inefficiently protecting existing industries against environmentally and economically preferable innovations and the roles of a wide range of other policies in encouraging or retarding competitive evolution of businesses toward more environmentally sustainable performance levels (not to mention the improvement of environmental performance of public sector business units themselves). These phenomena become topically acute as industries are restructured through changes such as deregulation (i.e., the utility industry).

On another level, new governance models are under investigation that will help mobilize private investment and innovation in environmental initiatives. Some are looking at the more recent phenomena of self-regulation (such as the Global Reporting Initiative or the Forest Stewardship Council) (Prakash, 2002). While maintaining a solid foundation of government regulation upon which to build new forms of innovative policies, others are looking to alternative regulatory programs that employ a negotiated form of compliance tailored to the needs and potentialities of individual organizations and environmental contexts. This new approach is "characterized by a new kind of legal self-restraint . . . [which] restricts itself to the installation, correction, and redefinition of democratic self-regulatory mechanisms" (Teubner, 1983:239). Cooperative environmental policy fundamentally reconfigures the role and objectives of both oversight agencies and the regulated community. Instead of mandating environmental policy, regulators seek out the input and participation of other parties with site-specific knowledge about the nature of environmental problems they encounter. Through the strategic steering of networks (DeBruijn and Heuvelhof, 2000), potentially innovative solutions are developed to resolve environmental problems. These may include regulated private sector organizations, non-profit organizations, scientific communities, local and state governments, community organizations, and others. Through negotiation among these interested parties, corporations gain the flexibility to define which emission sources to control through site-specific compliance strategies that achieve broadly defined objectives (Schmitt, 1994). Cooperative environmental policy strives to reward proactive companies for seeking competitive advantage through environmental innovation beyond regulatory standards (Fiorino, 1999). In that direction, the U.S. government has introduced a host of voluntary programs that are designed to foster collaboration between government agencies and regulated entities on the development of innovative, beyond-compliance environmental management solutions. The objective of such programs is compelling: to uncover ways for regulated entities to save money and achieve higher environmental protection standards than are guaranteed by existing regulations. Unfortunately, adoption of these programs has been slow. More research is necessary for understanding this new form of regulatory activity.

CONCLUSION

In today's business environment, annual costs for pollution control in the United States have risen nearly 600 percent since 1972, reaching levels equal to roughly 2 percent of the gross domestic product. As a result, companies are working on ways to devote resources toward environmental initiatives in a way that satisfies their economic objectives. They need a way to translate environmental issues into a form that they can understand and manage. Environmental protection, as an issue of corporate concern, has become much more complex and requires a more sophisticated view to be managed effectively. To treat environmental and business issues as separate and distinct leaves the business manager at a strategic disadvantage, unable to efficiently recognize the reality of a changing society—one that will demand ever greater corporate responsibility for protecting the environment. And this is an area where academic research can offer a contribution. But even more so, research into managerial decision making and the environment has implications for activists who now recognize that, to improve environmental conditions in today's world, they need to understand how to change the behavior of business; it also has implications for policy makers who need to understand how to incorporate business thinking into policy development so as to foster the most effective and efficient response from business.

And, in closing, it must be noted that research into corporate environmental behavior is now transitioning into new areas regarding sustainable development. The shift represents an expansion and augmentation rather than a change in focus within the research agenda. But although the concept of sustainable development has clearly entered the lexicon of corporate dialogue, the integration into business practice and research has far to go. Much research is needed in understanding how this concept will emerge, what it means, and where it is going. The existing and emerging research agenda on environmental issues has much to offer in shedding light on the triple bottom line of sustainable development: economic prosperity, environmental quality, and social equity (Elkington, 1998).

ACKNOWLEDGMENTS

The author would like to acknowledge and thank those who provided assistance on earlier drafts of this appendix: Brad Allenby, Richard Andrews, Garry Brewer, Bruce Clemens, Mary Gentile, Eric Hansen, Andy King, Ans Kolk, Ted London, Jennifer Nash, Ken Peattie, Steve Percy, Nigel Roome, Naomi Soderstrom, Ed Stafford, and Paul Stern. This work was supported by the Frederick A. and Barbara M. Erb Environmental Management Institute at the University of Michigan.

REFERENCES

Allison, G.
1971 *Essence of Decision: Explaining the Cuban Missile Crisis.* New York: Harper Collins Publishers.

Andreason, A.
1995 *Marketing Social Change: Changing Behavior to Promote Health, Social Development and the Environment.* San Francisco: Jossey-Bass.

Andrews, R.
1999 *Managing the Environment, Managing Ourselves: A History of American Environmental Policy.* New Haven, CT: Yale University Press.

Andrews, R., N. Darnall, D.R. Gallagher, S.T. Kevner, E. Feldman, M. Mitchell, D. Amaral, and J. Jacoby
2001 Environmental management systems: History, theory, and implementation research. Pp. 31-60 in *Regulating from the Inside: Can Environmental Management Systems Achieve Policy Goals?* C. Coglianese and J. Nash, eds. Washington, DC: Resources for the Future.

Aragón-Correa, J.A., and S. Sharma
2003 A contingent resource-based view of proactive corporate environmental strategy. *Academy of Management Review* 28(1):71-88.

Bansal, P., and K. Roth
2000 Why companies go green: A model of ecological responsiveness. *Academy of Management Journal* 43(4):717-736.

Barney, J.
1991 Firm resources and sustained competitive advantage. *Journal of Management* 17(1):99-120.

Baumol, W., and A. Blinder
1985 *Economics: Principles and Policy.* San Diego, CA: Harcourt Brace Jovanovich.

Beck, U.
1992 *Risk Society: Towards a New Modernity.* London, England: Sage.

Berman, S., A. Wicks, S. Kotha, and T. Jones
1999 Does stakeholder orientation matter? The relationship between stakeholder management models and firm financial performance. *Academy of Management Journal* 42:488-506.

Bloom, P., and G. Gundlach, eds.
2001 *Handbook of Marketing and Society.* Thousand Oaks, CA: Sage Publications.

Botkin, D.
2004 *Our Natural History: The Lessons of Lewis and Clark.* Oxford, England: Oxford University Press.

Buttel, F.
1987 New directions in environmental sociology. *Annual Review of Sociology* 13: 465-488.
1992 Environmentalism: Origins, processes, and implications for rural social change. *Rural Sociology* 57(1):1-27.

Catton, W., and R. Dunlap
1980 A new ecological paradigm for post-exuberant sociology. *American Behavioral Scientist* 20(1):15-47.

Clarke, S., and N. Roome
1999 Sustainable business: Learning action networks as organizational assets. *Business Strategy and the Environment* 8(5):296-310.

Clarkson, M.
1995 A stakeholder framework for analyzing and evaluating corporate social performance. *Academy of Management Review* 20:92-117.

Colby, M.
1991 Environmental management in development: The evolution of paradigms. *Ecological Economics* 3:193-213.

Cronon, W.
1991 *Nature's Metropolis: Chicago and the Great West.* New York: W.W. Norton & Company.
2003 *Changes in the Land: Indians, Colonists and the Ecology of New England.* New York: Hill and Wang.

Cropper, M., and W. Oates
1992 Environmental economics: A survey. *Journal of Economic Literature* 30:675-740.

DeBruijn, J., and E. Heuvelhof
2000 *Networks and Decision Making.* Utrecht, The Netherlands: Lemma.

Delmas, M.
2002 The diffusion of environmental management standards in Europe and in the United States: An institutional perspective. *Policy Sciences* 35(1):91-119.

Delmas, M., and A. Terlaak
2001 A framework for analyzing environmental voluntary agreements. *California Management Review* 43(3):44-63.

Dietz, T., E. Ostrom, and P. Stern
2003 The struggle to govern the commons. *Science* 302(12):1907-1912.

Dunlap, R., and W. Catton
1979 Environmental sociology. *Annual Review of Sociology* 5:243-273.
Easterbrook, G.
1995 *A Moment on the Earth*. New York: Viking.
Egri, C., and S. Herman
2000 Leadership in the North American environmental sector: Values, leadership styles, and contexts of environmental leaders and their organizations. *Academy of Management Journal* 43(4):571-604.
Eliot, R., and A. Gore, eds.
1983 *Environmental Philosophy: A Collection of Readings*. University Park, PA: Pennsylvania State University Press.
Elkington, J.
1998 *Cannibals with Forks: The Triple Bottom Line for 21st Century Business*. Oxford, England: Capstone.
Fiorino, D.
1999 Rethinking environmental regulation: Perspectives on law and governance. *Harvard Environmental Law Review* 23(2):441-469.
Fischer, K., and J. Schott, eds.
1993 *Environmental Strategies for Industry: International Perspectives on Research Needs and Policy Implications*. Washington, DC: Island Press.
Flannery, B., and D. May
2000 Environmental ethical decision making in the U.S. metal-finishing industry. *Academy of Management Journal* 43(4):642-662.
Fligstein, N.
1997 Social skill and institutional theory. *American Behavioral Scientist* 40(4):397-405.
Fombrun, C., and M. Shanley
1990 What's in a name? Reputation building and corporate strategy. *Academy of Management Journal* 33:233-258.
Friedman, R.
2000 When you find yourself in a hole, stop digging. *Journal of Industrial Ecology* 3(4):15-19.
Frosch, R., and N. Gallopoulos
1989 Strategies for manufacturing. *Scientific American* 144:144-152.
Galambos, L.
1970 The emerging organizational synthesis in American history. *Business History Review* 44(Autumn):279-290.
Gentile, M.
2002 *What Do We Teach When We Teach Social Impact Management?* (Business and Society Program Discussion Paper IV.) New York: Aspen Institute.
Gladwin, T., J. Kennelly, and T. Krause
1995 Shifting paradigms for sustainable development: Implications for management theory and research. *Academy of Management Review* 20(4):874-907.
Hahn, R., and R. Stavins
1991 Incentive-based environmental regulation: A new era from an old idea. *Ecology Law Quarterly* 18(1):1-42.
Hannan, M., and J. Freeman
1977 The population ecology of organizations. *American Journal of Sociology* 82: 929-964.
Hannigan, J.
1995 *Environmental Sociology: A Social Constructionist Perspective*. London, England: Routledge.

Hargrove, E.
1989 *Foundations of Environmental Ethics.* Englewood Cliffs, NJ: Prentice Hall.
Hart, S.
1995 A natural-resource based view of the firm. *Academy of Management Review* 20(4):986-1014.
Hays, S.
1998 The future of environmental regulation. Pp. 109-114 in *Explorations in Environmental History: Essays.* S. Hays and J. Tarr, eds. Pittsburgh, PA: University of Pittsburgh Press.
Hoban, T., and R. Brooks
1996 *Green Justice: The Environment and the Courts.* Boulder, CO: Westview Press.
Hoffman, A.
2001 *From Heresy to Dogma: An Institutional History of Corporate Environmentalism.* Stanford, CA: Stanford University Press.
2003 Linking social systems analysis to the industrial ecology framework. *Organization & Environment* 16(1):66-86.
Hoffman, A., and J. Ehrenfeld
1998 Corporate environmentalism, sustainability and management studies. Pp. 55-73 in *Environmental Strategies for Industry: The Future of Corporate Practice,* N. Roome, ed. Washington, DC: Island Press.
Hoffman, A., and W. Ocasio
2001 Not all events are attended equally: Toward a middle-range theory of industry attention to external events. *Organization Science* 12(4):414-434.
Holmes, R.
1988 *Environmental Ethics.* Philadelphia, PA: Temple University Press.
Howard-Grenville, J., and A. Hoffman
2003 The importance of cultural framing to the success of social initiatives in business. *Academy of Management Executive* 17(2):70-84.
Hunt, C., and E. Auster
1990 Proactive environmental management: Avoiding the toxic trap. *Sloan Management Review* 31(2):7-18.
Interfaith Center on Corporate Responsibility
2003 *Companies, Resolutions and Status: 2002-2003 Season.* Available: http://www.iccr.org/shareholder/proxy_book03/03statuschart.php [February 16, 2005].
Intergovernmental Panel on Climate Change
1990 *Climate Change: The IPCC Scientific Assessment.* Cambridge, England: Cambridge University Press.
Jennings, P., and P. Zandbergen
1995 Ecologically sustainable organizations: An institutional approach. *Academy of Management Review* 20(4):1015-1052.
John, R.
1997 Elaborations, revisions, dissent: Alfred D. Chandler, Jr.'s "The Visible Hand After Twenty Years." *Business History Review* 71(Summer):151-200.
Kassinis, G., and N. Vafeas
2002 Corporate boards and outside stakeholders as determinants of environmental litigation. *Strategic Management Journal* 23:399-415.
King, A., and M. Lenox
2000 Industry self-regulation without sanctions: The chemical industry's responsible care program. *Academy of Management Journal* 43(4):698-716.
2001 Does it really pay to be green: An empirical study of firm environmental and financial performance. *Journal of Industrial Ecology* 5(1):105-116.

Kolk, A.
2003 Trends in sustainability reporting by the Fortune global 250. *Business Strategy and the Environment* 12:279-291.
Kolk, A., and R. van Tulder
2004 Internationalization and environmental reporting: The green face of the world's leading multinationals. *Research in Global Strategic Management* 9:95-117.
Krupp, F.
1990 Win/win on the environmental front. *EPA Journal* 16(5):30-31.
Kumar Naj, A.
1990 Industrial switch: Some firms reduce pollution with clean manufacturing. *The Wall Street Journal* December 24:A1.
Laclau, E., and C. Mouffe
1985 *Hegemony and Socialist Strategy: Towards a Radical Democratic Politics.* London, England: Verso.
Lawrence, A., and D. Morell
1995 Leading-edge environmental management: Motivation, opportunity, resources and processes. Pp. 99-126 in *Research in Corporate Social Performance and Policy,* J. Post, ed. Greenwich, CT: JAI Press.
Lawrence, T.
1999 Industrial strategy. *Journal of Management* 25(2):161-188.
Lober, D.
1996 Evaluating the environmental performance of corporations. *Journal of Managerial Issues* 8(2):184-205.
London, T., and D. Rondinelli
2003 Partnerships for learning: Managing tensions in nonprofit organizations' alliances with corporations. *Stanford Social Innovation Review* 1(3):28-35.
Lovins, A., L. Hunter Lovins, and P. Hawken
1999 A road map for natural capitalism. *Harvard Business Review* 77(4):145-158.
Margolis, J., and J. Walsh
2001 *People and Profits? The Search for a Link Between a Company's Social and Financial Performance.* Mahwah, NJ: Lawrence Erlbaum Associates.
McAdam, D., J. McCarthy, M. Zald, P. Lange, R. Bates, E. Comisso, P. Hall, J. Migdal, and H. Milner
1996 *Comparative Perspectives on Social Movements: Political Opportunities, Mobilizing Structures, and Cultural Framings.* Cambridge, England: Cambridge University Press.
McGurty, E.
1997 From NIMBY to civil rights: The origins of the environmental justice movement. *Environmental History* 2:301-323.
Mohr, L., D. Eroglu, and P. Ellen
1998 The development and testing of a measure of skepticism toward environmental claims in marketers' communications. *Journal of Consumer Affairs* 32(1):30-55.
Morrison, C.
1991 *Managing Environmental Affairs: Corporate Practices in the U.S., Canada and Europe.* New York: The Conference Board.
Nash, J.
2002 Industry codes of practice: Emergence and evolution. Pp. 235-252 in National Research Council, *New Tools for Environmental Protection: Education, Information and Voluntary Measures.* Committee on the Human Dimensions of Global Change, T. Dietz and P. Stern, eds. Division of Behavioral and Social Sciences and Education. Washington, DC: National Academy Press.

Ottman, J.
1998 *Green Marketing: Opportunity for Innovation.* New York: McGraw-Hill.

Peattie, K.
1999 Trappings versus substance in the greening of marketing planning. *Journal of Strategic Marketing* 7(2):131-148.

Porter, M., and C. van der Linde
1995 Green and competitive: Ending the stalemate. *Harvard Business Review* 73(5): 120-134.

Prakash, A.
2002 Factors in firms and industries affecting the outcomes of voluntary measures. Pp. 303-310 in National Research Council, *New Tools for Environmental Protection: Education, Information, and Voluntary Measures*, Committee on the Human Dimensions of Global Change, T. Dietz and P.C. Stern, eds. Division of Behavioral and Social Sciences and Education. Washington, DC: National Academy Press.

Purser, R., C. Park, and A. Montuori
1995 Limits to anthropocentrism: Toward an ecocentric organization paradigm. *Academy of Management Review* 20(4):1053-1089.

Ramus, C., and U. Stegner
2000 The roles of supervisory support behaviors and environmental policy in employee "ecoinitiatives" at leading edge European companies. *Academy of Management Journal* 43(4):605-626.

Revesz, R.
1997 *Foundations of Environmental Law and Policy.* New York: Oxford University Press.

Rondinelli, D., and T. London
2003 How corporations and environmental groups cooperate: Assessing cross-sector alliances and collaborations. *Academy of Management Executive* 17(1):61-76.

Roome, N.
1998 Implications for management practice, education, and research. Pp. 259-276 in *Sustainability Strategies for Industry,* N. Roome, ed. Washington, DC: Island Press.

Rosen, C.
1995 Businessmen against pollution in late nineteenth century Chicago. *Business History Review* 71(Fall):351-397.
1997 Industrial ecology and the greening of business history. *Business and Economic History* 26:123-137.

Rosen, C., and C. Sellers
1999 The nature of the firm: Towards an ecocultural history of business. *Business History Review* 73(Winter):577-600.

Roy, M.
1991 Pollution prevention, organizational culture and social learning. *Environmental Law* 22:188-251.

Royston, M.
1979 *Pollution Prevention Pays.* London, England: Pergamon.
1980 Making pollution prevention pay. *Harvard Business Review* 58(6):6-27.

Sarokin, D., W. Muir, C. Miller, and S. Sperber
1985 *Cutting Chemical Wastes.* New York: INFORM.

Schlegelmilch, B., G. Bohlen, and A. Diamantopoulos
1996 The link between green purchasing decisions and measures of environmental consciousness. *European Marketing Journal* 30(5):35-55.

Schmidheiny, S.
1992 *Changing Course.* Cambridge, MA: MIT Press.

Schmitt, R.
1994 The AMOCO/EPA Yorktown experience and regulating the right thing. *Natural Resources and Environment* (Summer):11-13, 51.

Schnaiberg, A., and K. Gould
1994 *Environment and Society: The Enduring Conflict.* New York: Worth Publishing.

Sefcik, S., N. Soderstrom, and C. Stinson
1997 Accounting through green colored glasses: Teaching environmental accounting. *Issues in Accounting Education* 12(1):129-141.

Sexton, K., A. Marcus, W. Easter, and T. Burkhardt, eds.
1999 *Better Environmental Decisions: Strategies for Governments, Businesses and Communities.* Washington, DC: Island Press.

Sharma, S.
2000 Managerial interpretations and organizational context as predictors of corporate choice of environmental strategy. *Academy of Management Journal* 43(4):681-697.

Shrivastava, P.
1995 The role of corporations in achieving ecological sustainability. *Academy of Management Review* 20(4):936-960.

Smart, B.
1992 *Beyond Compliance.* Washington, DC: World Resources Institute.

Starik, M., and A. Marcus
2000 Introduction to the special research forum on the management of organizations in the natural environment: A field emerging from multiple paths, with many challenges ahead. *Academy of Management Journal* 43(4):539-546.

Starik, M., and G. Rands
1995 Weaving an integrated web: Multilevel and multisystem perspectives of ecologically sustainable organizations. *Academy of Management Review* 20(4):908-935.

Starke, L.
2004 *State of the World 2004: A Worldwatch Institute Report on Progress Toward a Sustainable Society.* New York: W.W. Norton & Company.

Stead, E., and J. Stead
1995 *Management for a Small Planet.* London, England: Sage Publications.

Teubner, G.
1983 Substantive and reflexive elements in modern law. *Law and Society Review* 17: 239-252.

Tietenberg, T.
1992 *Environmental and Natural Resource Economics.* New York: Harper Collins Publishers.
2002 The tradable permits approach to protecting the commons: What have we learned? Pp. 197-232 in National Research Council, *The Drama of the Commons,* Committee on the Human Dimensions of Global Change, T. Dietz, N. Dolsak, P.C. Stern, S. Stonich, and E.U. Weber, eds. Division of Behavioral and Social Sciences and Education. Washington, DC: National Academy Press.

Times Mirror
1995 *The Environmental Two Step: Looking Forward, Moving Backward.* New York: Times Mirror Magazines.

U.S. Environmental Protection Agency
2000 *Industrial Ecology and EPA: Report on the EPA Industrial Ecology Workshop.* Workshop Organizing Committee. Washington, DC: U.S. Environmental Protection Agency.

Weizacker, E., A. Lovins, and H. Lovins
1998 *Factor Four: Doubling Wealth—Halving Resource Use: A Report to the Club of Rome*. London, England: Earthscan.
Wernerfelt, B.
1989 A resource-based view of the firm. *Strategic Management Journal* 5(2):171-180.
World Business Council on Sustainable Development
1997 *Exploring Sustainable Development: WBCSD Global Scenario*. London, England: World Business Council on Sustainable Development.
World Resources Institute
1994 *World Resources, 1994-1995*. New York: Oxford University Press.
1999 *Critical Consumption Trends and Implications: Dedgrading Earth's Ecosystems*. Washington, DC: World Resources Institute.
2000a *Weight of Nations: Material Outflows from Industrial Economies*. Washington, DC: World Resources Institute.
2000b *World Resources, 2000-2001: People and Ecosystems*. Washington, DC: World Resources Institute.
2001 *Understanding the Forest Resources Assessment 2000*. Washington, DC: World Resources Institute.
2003 *World Resources 2002-2004: Decisions for the Earth: Balance, Voice, and Power*. Washington, DC: World Resources Institute.

Appendix D

Forecasting for Environmental Decision Making: Research Priorities

William Ascher

INTRODUCTION

This brief survey is intended to identify where research on forecasting for environmental decision making is most promising, rather than to assert best practices in the choices of methods and process. Nevertheless, some of the premises and dimensions should be clarified in order to support the recommendation in this analysis.

Premises

An assessment of the research needs for environmental forecasting should rest on three basic premises. First, decision needs (both short term to address today's policy challenges and long term to improve the scientific capacity to address future policy challenges) should drive the selection of forecasting foci, methodologies, and assessments. Therefore, it is important to set the research objectives for improving environmental forecasting according to the needs of formulating environmental policies and decisions. This does not mean that forecasting in order to improve science is less important than forecasting to meet immediate policy needs. It does mean that considerations of the short- and long-term usefulness of forecasting should drive the research agenda. This depends on aspects of the forecasts per se (reliability, credibility, completeness, and relevance to the policies and specific decisions) and on how the forecasting exercise interacts with the other facets of the decision-making process. Research on how to make forecasts more useful is as important as improving the accuracy of the forecasting methods.

Second, given the importance of utility, conveying the magnitude and nature of uncertainty is crucial. It is a central concern of research on the communication of scientific information. In addition, determining the magnitude and nature of uncertainty is an essential research task, as is the task of understanding how uncertainty affects the decision process.

Third, forecasting is essential regardless of the approach to environmental and resource management. Even if decision makers engage in what they regard as adaptive management,[1] forecasting is still required in the selection of optimal strategies. If feedback through monitoring and evaluation calls for policy changes, the decision makers still must project the likely outcomes of available alternatives; without this analysis the adaptation is just as likely to result in a deterioration of outcomes. If adaptive management resorts to policy experiments in the vein of Carl Walters, it is still essential to predict whether the outcomes pose unacceptable risks that would outweigh the benefits of learning through experimentation.

Preview of Needs

Sound environmental decision making requires forecasts that are

- more comprehensive in terms of input considerations, outcomes and effects
- sensitive to threshold effects (nonlinearities)
- better linked to *valuation* of outcomes and effects so that they can assist policy makers and the public to understand the magnitude of the costs, risks, and opportunities
- provide a strong sense of how people are affected
- perceived as credible[2] if credibility is deserved
- convey the degree (and nature) of their uncertainty, such that hedging strategies can be developed

For the forecasting effort (as distinct from the substantive forecast content) to make the most effective contribution to the decision process, it should

- engage decision makers in the process so that they can ensure the relevance of the choice of what is forecasted and gain confidence in the process
- focus decision makers' attention on emerging problems and opportunities
- provide adequate participation for stakeholders (although what is adequate depends on the specific property rights regime, legal mandates, and other contextual factors)

- involve sufficiently balanced sponsorship (in terms of funding, analytic effort, and review) to bolster its reliability and perceived credibility

How can research address these needs in the short and medium term (i.e., within the next five years or so)? One task is to inventory and assess the approaches and methodologies that already exist and give prominence to the sound alternatives. Because the forecasting task is frequently just one component of analysis, only a limited number of analysts specialize sufficiently in forecasting methodology to have the incentives or time to stay abreast of the forecasting literature, which is dispersed across many journals on forecasting, risk assessment, sectoral specialties (such as energy or land use), limited-circulation reports, and books. "Best practice" inventories are therefore very important, as long as analysts keep in mind that different questions may require different practices.

The other task is to develop new approaches where we are reasonably confident that existing approaches are inadequate *and* there is reason to believe that progress can be made. Yet some areas where one might think improvements can be made may be dead ends because of the intrinsic limitations of the forecasting task.

To think through where assessment and research are most needed, it is useful to distinguish 11 aspects of the forecast and the forecasting effort:

1. the units of analysis (e.g., disaggregated trends versus aggregate trends; impacts on particular groups versus national impacts)
2. the methodological approach of the forecasting effort (e.g., econometric models, systems dynamics models, scenario writing, extrapolation)
3. perceived appropriateness of the methods
4. the transparency of assumptions and methods
5. the theoretical content that drives the projections within the chosen methodology (e.g., the relationship between industrial expansion and pollution within an econometric or systems dynamic model)
6. the modes of expressing forecasts and the uncertainties of these forecasts
7. sponsorship
8. integration of the forecasting task with other decision processes (decision-maker involvement; stakeholder involvement)
9. reputation of the forecasters
10 breadth of the expertise of those involved in the forecasting effort
11. potential for the forecasting effort to contribute to the identification of additional policy options

APPROACHES TO ENHANCING THE ACCURACY AND RELIABILITY OF FORECASTS FOR ENVIRONMENTAL DECISION MAKING

Comprehensiveness of the Initial Mapping of Forecast-Relevant Factors

The accuracy and reliability of projections depend on sequencing and balancing comprehensiveness and selectivity.[3] The first challenge is to ensure that a sufficient range of potential influences is taken into account in a preliminary assessment so that relevant factors are not ignored. For example, technological progress inputs were often missing from long-term environmental models (this was a highly criticized shortcoming of the Club of Rome models). This broad initial mapping does not mean that all of the factors considered will warrant the same degree of analytic attention or inclusion in the models that ultimately drive the subsequent analysis. The selectivity that follows the initial mapping must reflect both the finite nature of analytic resources and, less obviously, the match between the methods and the understanding of the system. Highly complex models that include poorly understood factors run the dual risks of imparting greater error and making the assessment of uncertainty even more difficult.

The challenge of making forecasts more comprehensive in terms of the range of trends has been taken up by many forms: systems dynamics models, integrated scenario writing, and integrated assessment models. The question, then, is which of these approaches can combine multiple trends reliably without becoming black boxes of such complexity in their operations and outputs that forecast users cannot grasp the dynamics of the interactions and therefore cannot assess the model's coherence, reliability, or sensitivity to variants in the assumptions. A compilation and assessment of these approaches would be a very useful research project.

An especially important but often overlooked aspect of effects is the adjustment cost associated with shifting to new technologies to mitigate environmental problems and resource scarcities. It is easy to invoke new technologies as the answers to environmental risks—for example, replacing hydrocarbons to reduce pollution levels. Yet the costs of new infrastructure as well as direct equipment are often underestimated, as is the time needed to make these adjustments. It would be highly worthwhile to assess the existing methodologies for estimating mitigation and transition costs.

While many "integrated assessment models" try to provide policy makers with both the identification of trends that policy must address and the implications of specific policy choices, a common weakness of such models is the absence of modeling the myriad policy responses themselves, which influence the trend patterns that have to be addressed during later periods. The early Club of Rome models presumed no responses to greatly increased pollution, resource depletion, etc. Some efforts, particularly from economics, try to finesse the representation of policy response by using aggregate-level optimization models. For example, the policy response to higher energy prices might be modeled by assuming that adjustments will minimize the risk-adjusted energy costs, implying that transitions from one energy source to another follow a highly rational logic. However, the limitations of this approach lie in the strong possibility that real systems do not adjust automatically and immediately (due to rigidities in shifting policies and the uncertainties that policy makers face) and the fact that policy makers' preferences reflect institutional and personal interests that cannot be captured by assuming system-level optimization. It is worth assessing whether policy response models can be developed through historical or theoretical analyses of the conditions and tipping points of policy response, through scenario writing techniques, or through simulations in which individuals are asked to play out various policy-making roles.

Forecasting Nonlinear Trends

The capacity to model long-term effects is especially challenging, as the cumulative impacts of gradual changes are often subject to threshold effects that are difficult to model and time. Significant effort has been put into determining how threshold effects can be represented mathematically, but this does not yield insights as to what levels actually trigger nonlinear changes. Thresholds occur when there are changes in the interactions between drivers and affected aspects of the ecosystem (e.g., pollutant concentrations impinge on the chemical processes of life forms; depletion of particular resources makes them price uncompetitive for large-scale use, as in the case of certain timber and fish species). The forecasting approaches would have to be able to merge the trends in the drivers with knowledge of their biophysical and economic impacts. Considerable uncertainty will remain. Beyond this, it may be that the best we can do is sensitivity analysis to determine how much uncertainty is implied by the range of reasonable assumptions about biophysical and economic threshold levels.

Forecasting Rare Events

As with forecasting nonlinear trends, the forecast of rare events has been addressed in terms of mathematical representation, but the challenge of developing reliable methods for assessing the potential impact of the entire *set* of low-probability events has hardly been addressed (Cleaves, 1994). By their very nature, the exhaustiveness of the list of "surprises" can rarely be assured. Even for identified events, low-probability events are difficult to characterize in terms of magnitude (for example, a war would certainly affect the environment in various ways, but what scope of war to posit?). Research on developing methods for taking low-probability events into account is worthwhile because of the importance of the task if accomplished, but one should not be overly confident about the chances of success.

Improving Reliability Through Model Testing and Evaluation

The biggest obstacle to weeding out poor environmental forecasting techniques and models is the long time horizons required of most environmental forecasting efforts. The question, then, is whether other approaches to gauging reliability can be helpful. How effective are

- the tests of short-term forecasts of models developed to do long-term forecasting?
- assessments of the track record of particular approaches in predicting the outcomes that have already occurred?
- backcasting (i.e., evaluating the capacity of the method to "predict" the historical pattern)?
- comparisons of different models of apparently equivalent levels of expertise in order to show how much uncertainty exists within "the state of the art"?

Each of these approaches has some obvious limitations, and although each can help to identify efforts that already have shown indications of deficiency, "passing" these tests does not ensure reliability. Specifically, the short-term accuracy of simple extrapolations or growth curves, as well as more complex models, cannot speak to the possibility that unanticipated changes in patterns and parameters will emerge subsequently. The value of historical assessments of particular methods is limited by, first, the fact that some methods work particularly well in one period but poorly in another and, second, that the success of earlier versions of an approach may not reflect future success, as methods are subjected to what their developers consider to be continual improvement. In addition, many environmental forecasts are conditional projections, insofar as they specify policy responses (or the lack thereof) as premises rather than predictions, and these policy response conditions rarely hold precisely.[4] The value of backcasting is compromised by the fact that parameters are typically chosen on the basis of past patterns, as opposed to the possibility that future parameters may change greatly. Finally, although the discrepancies among forecasts by different state-of-the-art approaches reflect uncertainty, the lack of discrepancies does not necessarily reflect reliability and certainty. Nevertheless, it is worthwhile to assess whether combinations of these approaches can succeed in screening out problematic approaches and evaluating degrees of uncertainty.

APPROACHES TO ENHANCING THE USEFULNESS OF FORECASTING FOR ENVIRONMENTAL DECISION MAKING

The usefulness of forecasts for environmental decision making has five aspects worth analyzing: (1) the capacity to link the biophysical trends to general socioeconomic outcomes and effects that policy makers must address; (2) the capacity to identify specific impacts on particularly relevant groups (e.g., children, the elderly, and other vulnerable populations); (3) the perceived credibility of the forecasts so that they are more likely to be influential—when appropriate—in policy choice; (4) the appropriate identification and expression of uncertainty; and (5) the integration of the forecasting effort with the other facets of the decision-making process.

Biophysical Trends and Socioeconomic Outcomes and Effects: Forecasting and Valuation

The usefulness of forecasts depends on an additional facet of comprehensiveness: projecting outcomes and effects in addition to drivers. The decision process works by selecting options on the basis of projected outcomes and longer-term effects and then valuing these effects, taking uncertainty and risks into account. Environment forecasting tends to focus on the physical trends, with only very tentative and often crude methodologies for linking the physical trends to the socioeconomic ones. The decision process cannot digest drivers by themselves. For example, the forecast of a 3°C mean temperature increase, or twice the SO_2 concentration, is not useful for decision making without knowing the physical impacts and the socioeconomic consequences of these impacts. To improve this state of affairs, forecasting efforts, whether through formal modeling (integrated assessment models [see for example, van Asselt, 2000]) or organized judgmental techniques such as the Delphi, have to project outcomes (e.g., disease incidence, population movements, crop changes, etc.) and the economic costs or benefits of these trends. A strong assessment of these integrated assessment models was completed nearly a decade ago (Weyant et al., 1996); another should be undertaken to determine whether they are bringing enough policy-relevant outcomes and effects into the analysis.

The linkage between biophysical forecasting and socioeconomic outcomes and effects depends on combining forecasting and valuation, each a big challenge in itself. Valuation often requires a level of detail that the forecasts lack, sometimes for the good reason that the range of plausible outcomes does not permit such specificity. Long-term economic effects are particularly difficult to forecast, because the markets, tastes, and technology may change in highly unpredictable ways.[5] An inventory and assessment of studies that integrate forecasting and valuation would be useful for gauging how well current approaches work and for identifying the obstacles to improvement.

Improving the Sense of Impacts on People: Case-Wise Forecasting

By forecasting the impacts of environmental outcomes and effects on particular classes or types of individuals, rather than simply projecting aggregate trends, the significance of these impacts for policy making can be better assessed, and policy options can be targeted better as well. For example, the impact of increased industrial concentration on nonmobile retirees who are susceptible to emphysema is likely to be more policy relevant than the trends in average exposure. Thus case-wise projections can often provide more policy-relevant insights than forecasting aggregate trends (Brunner and Kathlene, 1989). Case-wise analysis entails defining a cluster of cases that represent a policy-relevant category, finding prototypes or "specimens." Yet developing methodologies of case-wise forecasting that are sufficiently comprehensive and credible remains incomplete.

Assessing Magnitudes and Types of Uncertainty

The degree of confidence in the accuracy of the forecasts plays a pivotal role in determining the degree to which hedging strategies are required. Waiting to eliminate all uncertainty is absurd, although it is often used as a political ploy for inaction. Rather, the decision process needs to recognize and factor in uncertainty. This raises four questions:

1. Do the expressions of uncertainty get heard accurately? For many types of environmentally related forecasts, the degree and nature of uncertainty is not communicated. For example, Tarko and Songchitruska (2003:2) point out that the U.S. Highway Capacity Manual and comparable manuals of other countries "return point estimates that in most cases are mean values . . . none of the existing capacity manuals handle uncertainty in their procedures."
2. Can protocols, or at least expectations, be established that require forecasters to report on their uncertainty in constructive ways? Placing such requirements in manuals is one approach, but it may be possible to establish protocols that would be considered as best practice if not compulsory, such that a forecaster who does not express uncertainty constructively would risk a reputational loss. Some progress has been made in this regard through the Moss and Schneider (2000) recommendations for consistent

reporting of uncertainty for authors of the Intergovernmental Panel on Climate Change.[6] Similarly, Weiss (2002) has proposed a code of ethics for presenting analysis with clear indications of what constitutes what is believed to be fact, "mainstream" opinion, minority opinion, etc., based on legal distinctions among categories of evidence.

3. Do the different sources of uncertainty get recognized as different—which is important for both credibility and for determining how to reduce or cope with uncertainty? The important point here is that understanding the nature of uncertainty in environmental forecasts has the double role of helping to make environmental decisions now and helping to refine the forecasting techniques themselves. Impressive claims have been made about the potential for partitioning uncertainty into various categories and developing cost-research research allocations on this basis (e.g., the Senior Hazardous Analysis Committee of the U.S. Nuclear Regulatory Commission [Budnitz et al., 1997]). Others have questioned the categories of uncertainty (National Research Council, 1997) and express some skepticism about the feasibility of this approach. Nonetheless, it would be worthwhile to apply the uncertainty-parsing techniques to a wide range of environmental forecasting efforts to determine how promising further elaborations of these techniques might be. It would also be worthwhile to develop a taxonomy of uncertainty that is more comprehensive and practical in terms of current decision and directions for improving the methodologies. The distinction between so-called epistemic uncertainty (incomplete knowledge about a phenomenon) and aleatory uncertainty (intrinsic randomness) has been shown to be inadequate. The classification of uncertainty as aleatory depends as much on the model employed as on the state of knowledge of the phenomenon. For example, earthquake prediction errors that arise because the models do not attempt to incorporate the details of specific faults would be regarded as due to aleatory uncertainty; the errors of models that try to model particular faults would be considered as epistemic uncertainty.[7]

4. Insofar as further work on environmental forecasting techniques is useful, can we identify the sources of uncertainty in order to focus efforts on improved theory, better information about the state of nature, better parameter estimation of the models, better exogenous time series, and so on? In short, fine distinctions among types of uncertainty, such as distinguishing between uncertainty about the laws of nature and uncertainty about the state of nature, can help orient the most useful research for reducing uncertainty. For example, gaining greater accuracy in projecting certain fish populations may require costly investment in monitoring ocean temperatures and existing fish stocks, but if the models are weak in understanding how fish stocks behave, it may be more cost-effective to improve the models than the monitoring.

Over the past 30 years, forecasting has made remarkable progress in abandoning the compulsion to project just "the most likely" trend. Quite sophisticated approaches to conducting and displaying sensitivity analyses have been developed (see, for example, Prinn et al., 1999). Yet significant challenges remain in identifying the sources of uncertainty and expressing the magnitude of uncertainty in ways that are most useful to policy making, especially in light of the huge complexity that these latest efforts introduce. Developments of pluralistic collaborations, in which the forecasts generated by multiple models or experts are displayed without forcing a consensus projection, have emerged during the past few years (Rotman and van Asselt, 2001). The purpose of pluralistic collaborations is not to force or even promote convergence. Nor—as mentioned above—do similar outcomes imply that the forecasts are more reliable, given the common state of affairs that environmental forecasters share outlooks, methods, and data. The appropriate purpose of pluralistic collaborations is to identify where the assumptions are contestable and to demonstrate the implications of different assumptions. Whether this demonstration comes through to stakeholders and decision makers, to help them hedge against uncertainty, remains to be evaluated.

Assessing the Reactions to Uncertainty

In light of the inevitability of some degree of uncertainty in environmental forecasting, it is important to understand how analysts, stakeholders, and decision makers react to uncertainty. For example, a common reaction by the public to the perception of uncertainty is to become more skeptical of scientific input; the expression of uncertainty violates the public's "view of science as a simple logical process producing unequivocal answers" (Collins and Bodmer, 1986:98). A common reaction of analysts engaged in ecosystem valuation is to downplay the less certain aspects of value, frequently leaving the more straightforwardly measurable and monetizable aspects to dominate the valuation, often neglecting such benefits as aesthetics and existence values. In some circumstances, uncertainty is seized upon by status-quo-favoring politicians to paralyze policy action. Through a better understanding of the reactions to uncertainty, better processes for coping with uncertainty can be formulated.

Forecast Credibility

A forecast that is not perceived as credible is of little use, no matter how accurate and enlightening it may be. How can environmental forecasts be conveyed so that the honest expression of uncertainty does not undermine credibility? Without such assurances, forecasters will continue to be tempted to understate uncertainty so as to avoid losing credibility. It is known that perceived usefulness (which presumes credibility) in the eyes of policy makers often depends on whether the forecast (or other expert input) corresponds to the potential user's preconceived beliefs, whether the decision maker is involved in the analytical effort, and whether the analysis provides results that are compatible with the policy questions that must be decided (Weiss, 1977). In terms of perceived credibility per se, systematic research on its correlates is in its infancy, although it would seem reasonable to hypothesize that appropriate credibility would depend on (1) the track record of prior forecasts associated with the same forecasting group[8]; (2) the credentials of the forecasters, both personally and in terms of the prestige of their institutional affiliations; (3) the perception of impartiality, based in part on the neutrality or balance of the sponsorship of their studies[9]; (4) the transparency and plausibility of assumptions and methods[10]; (5) plausibility of the forecasts[11]; (6) the perception of honesty in expressing uncertainty; and (7) involvement of decision makers and stakeholders in the forecasting process.[12] User surveys would be useful for determining the impact of such strategies as forecasting multiple scenarios, expressing forecasts in probabilistic terms, forecasting ranges rather than point estimates, cofunding of forecasting efforts by opposing groups, and so on.

Integrating the Forecasting Effort into the Overall Decision-Making Process

Although it has become accepted wisdom that stakeholder and decision-maker participation in the forecasting efforts often provides impressive benefits, the optimal means for doing this remain understudied. It is likely that the best ways to involve stakeholders and decision makers will vary considerably across different contexts. Yet it would still be worthwhile to gather and analyze more cases, in the vein of the Cash et al. (2002) study, to determine both the core lessons and the variations on success or failure.

Forecasts as Decision Aids

Despite disappointments in the progress of simulations as analytic tools, their potential for use as heuristic models remains. The development of simulations as part of the toolkit of policy dialogue is not rocket science, but it may be a very important step in bringing environmental consequences to the attention of policy makers. Many prior simulation models designed for direct interactions with policy makers were black boxes without particularly reliable or credible outputs. A new generation of decision simulation tools is emerging, but they still have the problem that their increasing complexity can obfuscate the most robust dynamics.[13] These decision tools require that both the assumptions and the uncertainties be made explicit. One of the great virtues of doing multiple computer runs, whether with one or more models, is that the uncertainties become obvious as the variations in inputs or model specifications produce different outcomes. An inventory and assessment of current model-based decision-aid ensembles would be useful, with support for refining the most promising of them a good investment.

SUMMARY OF ASSESSMENT AND RESEARCH PRIORITIES

1. Compile and assess approaches to balance comprehensiveness and selectivity, specifically
   a. approaches for mapping potentially relevant dynamics and then selecting among them for further analysis
   b. methods for coping with multiple low-probability events
2. Assess approaches designed to combine biophysical and socioeconomic effects, specifically
   a. methods for estimating mitigation and transition costs
   b. methods for linking biophysical trends to valuation
   c. methods for anticipating policy responses and linking these responses to subsequent trends
   d. integrated assessment models
3. Develop sensitivity analysis approaches to gauge the uncertainty levels of applications of nonlinear trend forecasting.
4. Develop more general approaches to case-wise analysis of environmental change impacts on specific groups and explore how to present case-wise results to enhance reliability and perceived credibility.
5. Develop and promote protocols for the systematic expression of uncertainty in forecast reports.
6. Assess whether the presentation of different results ("pluralistic collaborations") helps to convey uncertainty appropriately and to clarify the implications of different assumptions.
7. Assess the reactions to uncertainty on the part of the public, stakeholders, and decision makers; develop processes that reduce the unconstructive reactions to uncertainty.
8. Refine the identification of correlates of perceived credibility (e.g., specific modes of decision maker participation; specific ways of expressing uncertainty).
9. Develop more refined frameworks to characterize types of uncertainty; assess whether identifying different types of uncertainty can help to make the uncertainty reduction more efficient.
10. Assess forecast evaluation techniques (short-term validation, track record of prior forecasts, backcasting, comparisons of alternative forecasts).

11. Assess existing methods and develop more refined approaches for stakeholder and decision-maker participation in environmental forecasting efforts.

12. Assess decision-aid forecasting models.

NOTES

1. The variation in adaptive management approaches is reflected in the contrast between Kai Lee's conception of adaptation as reevaluating optimal policy in reaction to feedback (Lee, 1993), and Carl Walters's conception of adaptation through ambitious experimentation, often nonoptimal, to learn enough about the system in order to formulate better policies (Walters, 1986).

2. "Credibility" has two senses: whether information is perceived as warranting acceptance as reliable and whether information intrinsically deserves to be perceived as such. To avoid confusion, this analysis will refer to perceived credibility and will refer to the latter sense as "reliability."

3. Lasswell (1971:86-88) lists comprehensiveness and selectivity, along with openness, dependability, and creativity, as the criteria for evaluating the intelligence function.

4. For example, a global warming forecast may be designed to project the consequences of the failure to enact the Kyoto Protocol. If the protocol is implemented more fully than this premise envisions, then the discrepancies between the predictions and the actual results are not "forecast errors" in the same sense as the discrepancies between an absolute forecast and actual outcomes. For the implications of conditional forecasting on the difficulty on assessing forecast accuracy, see Ascher (1989).

5. An interesting example of this can be found in Kenny et al. (2000), which tries to assess the economic impact of higher temperatures in New Zealand on the production of kiwifruit and corn, as well as the incursion of an invasive grass into fairy pasture. The economic consequences of higher temperatures on the kiwi crop are complicated not only by the uncertainties of temperature increases, but also whether technologies such as chemicals to inhibit premature budding and flowering of kiwi will progress, and whether the kiwi exports will remain as profitable over the long run given changes in tastes and potential competition from other countries.

6. The summary volume of *Climate Change 2001: Impacts, Adaptation, and Vulnerability* (McCarthy, Canziani, Leary, Dokken, and White, 2002) notes that among Intergovernmental Panel on Climate Change reports "there was no consistent use of terms to characterize levels of confidence in particular outcomes or common methods for aggregating many individual judgments into a single collective assessment. Recognition of this shortcoming . . . led to preparation of a guidance paper on uncertainties . . . for use by all . . . Working Groups and which has been widely reviewed and debated" (McCarthy et al., 2002:128).

7. The 1997 National Research Council panel that reviewed the Nuclear Regulatory Commission's *Probabilistic Seismic Hazard Analysis: Guidance on Uncertainty and Use of Expert* noted "epistemic uncertainty would be much greater if, in the assessment of seismic hazard at an eastern U.S. site, instead of representing random seismicity through homogeneous Poisson sources one used a model with an uncertain number of faults, each with an uncertain location, orientation, extent, state of stress, distribution of asperities, and so forth. As little is known about such faults, the total uncertainty about future seismicity and the calculated mean hazard curves would be about the same, irrespective of which model is used. However, the amount of epistemic uncertainty would be markedly different; it would be much greater for the more detailed, fault-based model. Consequently, the fractile hazard curves that represent epistemic uncertainty would also differ greatly . . . **[U]nless one accepts**

that all uncertainty is fundamentally epistemic, the classification of . . . uncertainty as aleatory or epistemic is ambiguous (National Research Council, 1997:32-33) (bold text in original).

8. Agrawala and Broad (2001:465) point to this factor in their assessment of climate change forecasts.

9. Busenberg (1999) found this in his assessment of scientific input on environmental risks such as oil spills.

10. For example, Gibbons (1999) notes the problems that nontransparency has caused in the credibility of predictions of the impact of genetically altered organisms.

11. This is closely related to the correspondence of the prediction with preconceived expectations (Weiss, 1977), but plausibility of previously unexpected predictions can be reinforced by explanation.

12. Weiss (1977) found this to be crucial to the impact of expert input in general in her study of U.S. federal executives' acceptance of technical input. Andrews (2002) found similar patterns in exploring the perceived legitimacy of scientific input by such organizations as the U.S. Office of Technology Assessment. Cash et al. (2002) found the interaction across the science-nonscience boundary to be important across a broad range of environmental issues.

13. An assessment of the usefulness of stakeholder group exposure to seven global climate change models concluded that "computer models were successful at conveying to participants the temporal and spatial scale of climate change, the complexity of the system and the uncertainties in our understanding of it. However, most participants felt that . . . most models were not sufficiently user-friendly and transparent for being accessed in an [Integrated Assessment] focus group" (Dahinden, Querol, Jäger, and Nilsson, 2000:253). Welp (2001:538) reaches the same conclusion.

REFERENCES

Agrawala, S., and K. Broad

2001 Integrating climate forecasts and societal decision making: Challenges to emergent boundary organizations. *Science, Technology, and Human Values* 26(4):454-477.

Andrews, C.J.

2002 *Humble Analysis: The Practice of Joint Fact-Finding.* Westport, CT: Praeger.

Ascher, W.

1989 Beyond accuracy: Progress and appraisal in long-range political-economic forecasting. *International Journal of Forecasting* 5(4):469-484.

Brunner, R.D., and L. Kathlene

1989 Data utilization through case-wise analysis: Some key interactions. *Knowledge in Society* 2:16-38.

Budnitz, R.J., G. Apostolakis, D.M. Boore, L.S. Cluff, K.J. Coppersmith, C.A. Cornell, and P.A. Morris

1997 *Recommendations for Probabilistic Seismic Hazard Analysis: Guidance on Uncertainty and the Use of Experts.* Senior Seismic Hazard Analysis Committee. Washington, DC: U.S. Nuclear Regulatory Commission.

Busenberg, G.

1999 Collaborative and adversarial analysis in environmental policy. *Policy Sciences* 32(1):1-11.

Cash, D.W., W. Clark, F. Alcock, N. Dickson, N. Eckley, and J. Jäger

2002 *Salience, Credibility, Legitimacy and Boundaries: Linking Research, Assessment and Decision Making.* (Working Paper RWP02-046.) Cambridge, MA: John F. Kennedy School of Government Faculty Research, Harvard University.

Cleaves, D.A.
1994 *Assessing Uncertainty in Expert Judgments About Natural Resources.* (U.S. Forest Service General Technical Report SO-1.) New Orleans, LA: U.S. Forest Service.
Collins, P.M.D., and W.F. Bodmer
1986 The public understanding of science. *Studies in Science Education* 13:98.
Dahinden, U., C. Querol, J. Jäger, and M. Nilsson
2000 Exploring the use of computer models in participatory integrated assessment—Experiences and recommendations for further steps. *Integrated Assessment* 1: 253-266.
Gibbons, M.
1999 Science's new social contract with society. *Nature* 402:C81-C84.
Kenny, G.J., R.A. Warrick, B.D. Campbell, G.C. Sims, M. Camilleri, P.D. Jamieson, N.D. Mitchell, H.G. McPherson, and M.J. Salinger
2000 Investigating climate change impacts and thresholds: An application of the CLIMPACTS integrated assessment model for New Zealand agriculture. *Climatic Change* 46:91-113.
Lasswell, H.D.
1971 *A Pre-View of Policy Sciences.* New York: Elsevier.
Lee, K.
1993 *Compass and Gyroscope: Integrating Science and Politics for the Environment.* Washington, DC: Island Press.
McCarthy, J.J., O.F. Canziani, N.A. Leary, D.J. Dokken, and K.S. White, eds.
2002 *Climate Change 2001: Impacts, Adaptation, and Vulnerability.* Cambridge, England: Cambridge University Press.
Moss, R.H., and S.H. Schneider
2000 Uncertainties in the IPCC TAR: Recommendations to lead authors for more consistent assessment and reporting. Pp. 33-51 in *Guidance Papers on the Cross Cutting Issues of the Third Assessment Report of the IPCC.* R. Pachauri, T. Taniguchi, and K. Tanaka, eds. Geneva, Switzerland: World Meteorological Organization.
National Research Council
1997 *Review of Recommendations for Probabilistic Seismic Hazard Analysis: Guidance on Uncertainty and Use of Experts.* Panel on Seismic Hazard Evaluation. Committee on Seismology. Board on Earth Sciences and Resources. Commission on Geosciences, Environment, and Resources. Washington, DC: National Academy Press.
Prinn, R., H. Jacoby, A. Sokolov, C. Wang, X. Xiao, Z. yang, R. Eckhaus, P. Stone, D. Ellerman, J. Melillo, J. FitzMaurice, D. Kicklighter, G. Holian, and Y. Liu
1999 Integrated global system model for climate policy assessment: Feedbacks and sensitivity studies. *Climatic Change* 41:469-546.
Rotman, J., and M. van Asselt
2001 Uncertainty management in integrated assessment modeling: Towards a pluralistic approach. *Environmental Monitoring and Assessment* 69:101-130.
Tarko, A.P., and P. Songchitruksa
2003 *Reporting Uncertainty in the Highway Capacity Manual—Survey Results,* Paper presented at the Transportation Research Board Annual Meeting, January 13, Washington, DC.
van Asselt, M.
2000 *Perspectives on Uncertainty and Risk.* Dordrecht, The Netherlands: Kluwer.
Walters, C.
1986 *Adaptive Management of Renewable Resources.* Caldwell, NJ: Blackburn Press.
Weiss, C.
2002 Scientific uncertainty in advising and advocacy. *Technology in Society* 24:375-386.

Weiss, C.H., ed.
1977 *Using Social Research in Public Policy Making*. Lexington, MA: D.C. Heath and Company.

Welp, M.
2001 The use of decision support tools in participatory river basin management. *Physics and Chemistry of the Earth, Part B: Hydrology, Oceans & Atmosphere* 26(7-8):535-539.

Weyant, J., O. Davidson, H. Dowlabadi, J. Edmonds, M. Grubb, E.A. Parson, R. Richels, J. Rotmans, P.R. Shakla, R.S.J. Tol, W. Cline, and S. Frankhauser
1996 Integrated assessment of climate change: An overview and comparison of approaches and results. Pp. 367-396 in *Climate Change 1995: Economics and Social Dimensions of Climate Change: Scientific-Technical Analysis*, Intergovernmental Panel on Climate Change (IPCC). Cambridge, England: Cambridge University Press.

Appendix E

Program Evaluation of Environmental Policies: Toward Evidence-Based Decision Making

Cary Coglianese and Lori D. Snyder Bennear

Do environmental policies work? Although this question is simple and straightforward, for most environmental policies it lacks a solid answer. This is not because no answers are available. On the contrary, there are often an abundance of purported answers to be found—just a shortage of systematic, empirical support for these answers. Decision making over environmental policy has too often proceeded simply on the basis of trial and error, without adequate or systematic learning from either the trials or errors. Decision makers often lack carefully collected evidence about what policies have accomplished in the past in order to inform deliberations about what new policies might accomplish in the future.

Obtaining systematic answers to the question of whether environmental policies work is vital. Any environmental policy should make a difference in the world, ideally changing environmental conditions for the better or at least preventing them from getting worse. Although intuitions and anecdotes may provide some reason for suspecting that a given policy has made or will make a difference, the only way to be confident in such suspicions is to evaluate a policy's impact in practice. Program evaluation research provides the means by which analysts can determine with confidence what works, and what does not, in the field of environmental policy. The results of program evaluation research can then be used by others when deciding if they should retain existing policies or adopt new or modified ones.

Although important program evaluation research has examined the impact of some environmental policies, such research has been remarkably scarce relative to the overall volume of environmental policy decisions made at the state and federal level, as well as relative to the amount of evaluation research found in other fields, such as medicine, education, or transportation safety. A renewed and greatly expanded commitment to program evaluation of environmental policy would help move environmental decision making closer to an evidence-based practice.

In this paper, we begin by defining the role that empirical analysis can play in policy deliberation and decision making, distinguishing program evaluation research from other types of analysis, including risk assessment, cost-effectiveness analysis, and cost-benefit analysis. Although reliance on these other types of analysis has greatly expanded over the past several decades, most other forms of analysis take place before decisions are made; relatively little analysis takes place after decisions have been made and implemented, which is when program evaluation occurs. We argue that any policy process that takes analysis and deliberation seriously *before* decisions are made should also take seriously the need for research *after* decisions are made.

We next explain the kinds of methodological practices that program evaluation researchers should use to isolate the *causal* effect of a particular regulation or other policy initiative, that is, the change in outcomes that would not have occurred *but for* the program. Even if an environmental policy is correlated with a particular environmental or social outcome, this does not necessarily mean that there is a causal relationship between the policy initiative and the change in outcomes. Only by adhering to the type of methods we highlight here will researchers be able to isolate the effects of specific policy interventions and thereby inform environmental decision making.

Finally, we suggest that the present time is an especially ripe one for expanding program evaluations of environmental policies. Although program evaluation techniques have been available for decades and have certainly been advocated for use in the field of environmental policy, recent developments in policy innovation, government management, and data availability make the present time more conducive for an expanded program evaluation research agenda. During the past several decades, the U.S. Environmental Protection Agency (EPA) and the states have developed a variety of new approaches to environmental protection that are now ready for evaluation. The prevailing policy climate generally supports evaluation of government performance, as evidenced by the Office of Management and Budget's new Program Assessment Rating Tool and legislation like the Government Performance and Results Act. Moreover, given the increasing ease of access to data made possible by the Internet, researchers will find it easier today to expand program evaluation in the field of environmental policy. Evidence-based deliberation and decision making over environmental policy are probably closer to becoming routine practices today than they have ever been before.

THE ROLE OF PROGRAM EVALUATION IN ENVIRONMENTAL POLICY

Since the overarching purpose behind environmental policies is to improve environmental conditions, and often thereby to improve human health, program evaluation can identify whether specific policies are serving their purposes and are having other kinds of effects, such as reducing environmental inequities, imposing economic costs, or promoting or inhibiting technological change. In this section we show how program evaluation research fits into the policy process and serves an important role in environmental decision making.[1]

Environmental Policy Making and Implementation

The policy process begins with the recognition of a potential environmental problem and a response by the policy maker, often the legislature (Brewer and deLeon, 1983). The response typically takes the form of a statute imposing requirements on industry or delegating authority to a regulatory agency, like the EPA or Fish and Wildlife Service, to create specific requirements that industry must follow or develop other programs to achieve legislative goals. Legislation is then implemented by federal, state, or local regulatory agencies. Implementation often requires these agencies to establish additional, more specific mandates. At the federal level, for example, environmental and natural resources agencies promulgate hundreds of new regulations each year. These regulations typically fill in gaps about the precise level of environmental protection to be achieved, the type of policy instruments to use to achieve statutory goals, and the time frame for compliance with new regulations.

Policy implementation includes other kinds of choices as well. It can include education, licensing, and grant programs. It also can include the selection of enforcement or other strategies to ensure compliance with policies. Regulatory agencies must make decisions about how they will target firms for enforcement: randomly, in reaction to complaints, based on past history, based on size or other criteria related to the regulatory problem to be solved, or some combination of these or other factors. Moreover, agency inspectors can be instructed to approach their work in an adversarial manner—that is, going "by the book" and issuing citations for any violations found—or in a more cooperative manner whereby regulatory inspectors work with regulated entities to solve problems (Bardach and Kagan, 1982; Scholz, 1984; Hutter, 1989).

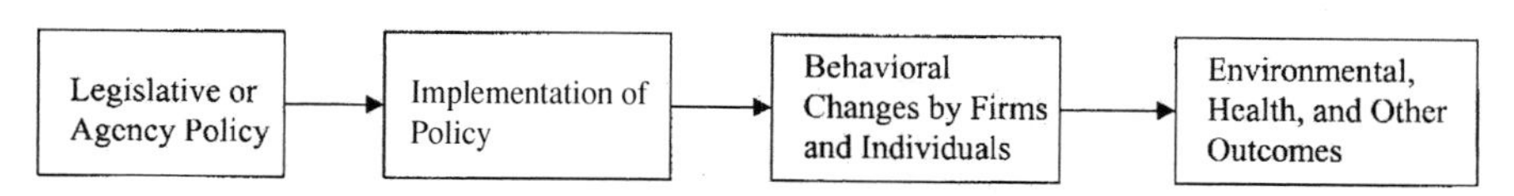

FIGURE E-1 A simple model of the environmental policy process.

Regulatory policies are adopted, and then implemented and enforced, in order to change the behavior of a class of businesses or individuals. The ultimate aim of policy making and implementation is to create incentives for individuals and firms to change their behavior in ways that will solve the problems that motivated the adoption of public policy in the first place. If a policy works properly, the behavioral change it induces will in turn result in the desired changes in environmental conditions, public health, or other outcomes. A basic diagram of the environmental policy process is provided in Figure E-1.

Prospective Analysis of Environmental Policy

Empirical analysis can usefully inform several stages of the policy process. During both the policy making and the implementation stages, analysis can inform deliberation and decision making about whether anything should be done to address an environmental problem and, if so, what set of policy instruments or strategies should be used. Currently, there are several different analytical methods used extensively during both policy making and implementation, including risk assessment, cost-effectiveness analysis, and benefit-cost analysis (Stokey and Zeckhauser, 1978). Each of these types of analysis is used prospectively to inform the deliberative process leading up to policy decisions.

Risk assessment characterizes the health or ecological risks associated with exposure to pollution or other hazardous environmental substances or conditions (National Research Council, 1983). It seeks to identify the causal relationships between exposure to specific environmental hazards and specific health or ecological conditions. As such, risk assessment seeks to provide a scientific basis for understanding the potential range of benefits that can be attained from policies that aim to reduce exposure to environmental hazards.[2]

Benefit-cost analysis seeks to help policy makers identify both the benefits and the costs of specific environmental policies and implementation strategies. It compares different policy or implementation alternatives based on their net benefits—that is, total benefits minus total costs (Arrow et al., 1996). Such analysis is usually conducted in advance of policy making to try to identify regulatory options that will be the most efficient (Viscusi, 1996; Hahn, 1998). As such, benefit-cost analysis usually leads to estimates of *expected* net benefits from different alternatives.

Cost-effectiveness analysis seeks to identify the lowest-cost means of achieving a specific goal (U.S. Environmental Protection Agency, 2000). Unlike benefit-cost analysis that compares alternatives in terms of both costs and benefits, cost-effectiveness analysis compares alternatives simply in terms of how much they cost in order to achieve a given goal—regardless of whether there will be positive net benefits from achieving this goal. For example, imagine that policy makers seek to reduce carbon dioxide emissions by 20 percent and that several policies could be selected that would achieve this desired level of reduction. Regardless of whether the 20 percent reduction maximizes net benefits, cost-effectiveness analysis can be used to help ensure the lowest-cost means to attain the selected goal.

Economic analyses of costs and benefits, along with risk assessments, are typically used prospectively in the regulatory process, that is, before government officials make decisions. The prospective use of these analytic techniques has expanded greatly in the past 20 years due to evolving professional practices as well as executive orders mandating economic analysis preceding the adoption of new federal regulations that are anticipated to impose $100 million or more in annual compliance costs (Coglianese, 2002; Hahn and Sunstein, 2002). These executive orders have existed under every administration since Ronald Reagan, and government agencies have developed detailed guidance for conducting the required analyses (U.S. Environmental Protection Agency, 2000; U.S. Office of Management and Budget, 2003a).

Retrospective Analysis: Program Evaluation of Environmental Policy

In contrast to the prospective role played by risk assessment and benefit-cost analysis, program evaluation occurs retrospectively, as it seeks to determine the impact of a chosen policy or implementation strategy after it has been adopted. For example, Snyder (2004a) evaluated the impact of pollution prevention planning laws that 14 states adopted in the 1990s. These laws required industrial facilities using toxic chemicals to develop plans for reducing their use of these chemicals. By forcing facilities to plan, these laws were supposed to encourage industry to find opportunities to lower their production costs as well as improve environmental protection. But did they work? Drawing on more than a decade's worth of data on toxic chemical releases by manufacturing plants in states with and without the planning laws, Snyder (2004a) found that the pollution planning laws had a measurable impact on plants' environmental performance. The planning laws were associated with a roughly 30 percent decline in releases of toxic chemicals.

Other regulatory policies have been evaluated retrospectively, including hazardous waste cleanup laws (Hamilton and Viscusi, 1999; Revesz and Stewart, 1995), air pollution and other media-specific environmental regulations (U.S. Environmental Protection Agency, 1997; Davies and Mazurek, 1998; Harrington et al., 2000; Chay and Greenstone, 2003), and information disclosure requirements, such as the Toxics Release Inventory (TRI) (Hamilton, 1995; Konar and Cohen, 1997; Khanna et al., 1998; Bui and Mayer, 2003). A variety of innovations in environmental policy have also received retrospective study, including market-based instruments (Stavins, 1998), voluntary programs (Alberini and Segerson, 2002; Arora and Cason, 1995, 1996; Khanna and Damon, 1999), and regulatory contracting programs like EPA's Project XL (Blackman and Mazurek, 2000; Marcus, Geffen, and Sexton, 2002). In addition, various rocedural "policies" have been subject to retrospective evaluation, such as the use of benefit-cost analysis (Morgenstern, 1997; Farrow, 2000; Hahn and Dudley, 2004) and negotiated rule making (Coglianese, 1997, 2001; Langbein and Kerwin, 2000). Finally, researchers have evaluated the impact of various types of enforcement strategies (Shimshack and Ward, 2003; May and Winter, 2000).

Like the Snyder (2004a) study, such retrospective analyses have sought to ascertain what outcomes specific policies have actually achieved.[3] Some of these outcomes are the ones the policy was intended to achieve, such as improvements in human health or the biodiversity of an ecosystem. However, program evaluation research also considers other effects, such as whether a policy has had unintended or undesirable consequences. Has it contributed to other problems similar or related to the one the policy was supposed to solve? What kinds of costs has the policy imposed? How are the costs and benefits of the policy distributed across different groups in society? Finally, program evaluation research can also focus on other outcomes including transparency, equity, intrusiveness, technological change, public acceptability, and conflict avoidance, to name a few.

By assessing the performance of environmental policies in terms of various kinds of impacts, retrospective evaluations can inform policy deliberations. Policy makers revisit regulatory standards periodically, sometimes at regular intervals specified in statutes or whenever industry or environmental groups petition for changes. More frequently, existing policies will be used as model solutions for new environmental problems, and so program evaluation of existing policies informs decisions about what policies to use in new situations. For this reason, program evaluation will also provide critical information for prospective analysis of new policy initiatives. By knowing what policies have accomplished in other contexts, pro-

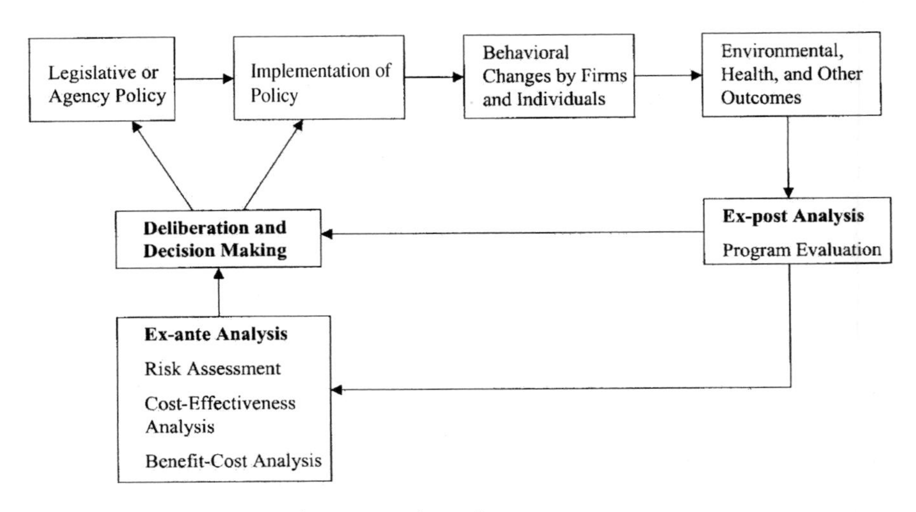

FIGURE E-2 Program evaluation in the policy process.

spective analyses—such as benefit-cost analysis—can be grounded in experience as well as theory and forecasting. The accuracy of the estimation strategies used in prospective analyses can also be refined by comparing ex ante estimates with the ex post outcomes indicated in program evaluations. Figure E-2 illustrates the role of program evaluation in the policy process.

METHODS OF PROGRAM EVALUATION

The goal of program evaluation is to ascertain the causal effect of a "treatment" on one or more "outcomes." In the field of environmental policy the treatment will often include government-mandated regulations that take the form of a range of policy instruments (Harrington, Morgenstern, and Sterner, 2004; Hahn, 1998). These regulations include technology and performance standards (Coglianese, Nash, and Olmstead, 2003), market-based instruments like emissions trading (Stavins, 2003), information disclosure policies (Kleindorfer and Orts, 1998), and management-based policies such as those requiring firms to develop pollution prevention plans (Coglianese and Lazer, 2003). The treatment could also consist of a variety of implementation strategies, ranging from different types of enforcement strategies, grant requirements, or public recognition and waiver programs, including such innovations as the EPA's Project XL, the National Environmental Performance Track, and the U.S. Department of the Interior's Habitat Conservation Plans (de Bruijn and Norberg-Bohm, 2001). The treatment could even include international treaties and nongovernmental initiatives that are designed to effect the environment, such as trade association self-regulatory efforts like the chemical industry's Responsible Care program or the wood and paper industry's Sustainable Forestry Initiative.

For each treatment to be evaluated, the researcher must obtain reliable measures of outcomes. Outcome measures used in evaluations of environmental policies can include measures of facility or firm environmental performance (e.g., emissions of pollutants, energy use), human health impacts (e.g., days of illness, mortality or morbidity rates), or overall environmental impacts (e.g., acres of wetland, ambient air quality). When the ultimate outcome of concern cannot be directly measured, proxies must be used to assess the impact of a policy. For example, it sometimes is not possible to assess an environmental policy in terms of its impact on reductions in human health risk, but researchers can use measures of pollution reduction as a proxy for the ultimate outcome of risk reduction.

Isolating the Causal Effects of Treatments on Outcomes

The goal of program evaluation is to go beyond simple correlation to estimate the causal effect of the treatment on the outcomes selected for study. A treatment and outcome may be correlated, but the treatment has "worked" only if it has had a causal effect on the outcome. To see how a researcher isolates the causal effect of one policy from all of the other potential explanations for a given change in the outcome, consider a hypothetical government program designed to encourage plant participation in a voluntary program that offers firms incentives for reducing pollution to levels below those needed to comply with existing regulations. The treatment is participation in the program and the outcome measure consists of emissions of pollutants from industrial facilities. In an ideal world, the researcher would observe the level of pollution each facility emits when it does not participate in the voluntary program. Then the researcher—again in an ideal (and imaginary) world—would travel back in time, assign each facility to participate in the program while leaving all other features of the facility unchanged, and observe the level of pollution each produces after it has participated in the program. If the researcher could actually observe, for each facility, both *potential* outcomes (that is, the outcome with and without treatment), then the causal effect of the program would be a straightforward difference between the pollution levels with and without participation.

Of course, the fundamental problem of causal inference is that researchers cannot travel back in time and reassign facilities from one group to another. In reality, researchers never observe both potential outcomes for any individual plant. They observe only the pollution levels of participating facilities, given that they participated, and the pollution levels for nonparticipating facilities, given that they did not participate. The challenge for program evaluation research is to use *observable* data to obtain valid estimates of the inherently *unobservable* difference in potential outcomes between the treatment and nontreatment (or control) groups.

Methods for Drawing Causal Inferences

How can researchers meet this fundamental challenge and draw reliable inferences about the causal effects of environmental policies?[4] If possible, the best approach would be to conduct a policy experiment and rely on random assignment of the treatment. If regulated entities subject to a treatment are assigned at random, then other factors that determine potential outcomes are also likely to be randomly distributed between the treatment and the control group. For example, with random assignment, there should not be any systematic differences in the treatment and control groups in terms of such things as industry characteristics, size of firms, or publicly traded versus privately held ownership. In the case of random assignment, any differences in outcomes between the two groups of entities can be attributed to the treatment.

True random experimental designs are, of course, rare or nonexistent in environmental policy. Regulation, voluntary program participation, and other treatments of interest are not generally randomly assigned. Instead, regulatory status is frequently determined by factors that are also correlated with potential outcomes such as the size of the facility, the facilities' pollution levels, the age of the facility, and so forth. For environmental policy analysis, researchers will generally be forced to use observational study designs—also referred to as quasi-experimental designs.[5] Observational studies do not rely on explicit randomization, rather they capitalize on "natural" treatment assignments (as a result these studies are also sometimes referred to as natural experiments). Because assignment to treatment is not random in observational studies, and treatment can be correlated with other determinants of potential outcomes, more sophisticated methods are required to isolate the causal effect of the treatment.

In observational studies where strict random assignment does not hold, there may be random assignment *conditional* on other observable variables. For example, imagine that one state's legislature passes a new regulation on hazardous waste while another state's does not. If the two states were quite similar—that is, they had the same types of facilities and the same socioeconomic and demographic variables—then the conditions of random assignment may be effectively met. If the states are not identical (that is, there are some differences in the types of facilities or community demographics), then observed differences in environmental performance across the states may be due to the difference in regulation or to the differences in these other variables. One state, for instance, may simply have larger or older industrial facilities that will affect how much hazardous waste they produce.

Variables that are correlated with the treatment and also with outcomes are called confounders—the presence of these variables confounds researchers' ability to draw causal inferences from a simple difference in average outcomes. If the confounders can be quantified with available data, however, then they are "observable."[6] If all of the confounders are observable, then the causal effect of regulation could be estimated by examining the difference in outcomes, conditional on the confounding variables. In our hypothetical two-state example, a researcher could estimate the causal effect of the treatment by controlling for confounders such as the size or age of the facilities in both states. The researcher would essentially be comparing the environmental performance of facilities in the two states that have the same size, age, and other characteristics related to the generation of hazardous wastes.

Program evaluation researchers find analytic techniques such as regression and matching estimators to be useful when conditional random assignment holds. Regression analysis estimates a relationship between the outcome measure and a set of variables that may explain or be related to the outcome. One of these explanatory variables is the treatment variable, and the others are the confounders (also called control variables). Regression analysis isolates statistically the effect of the treatment holding all of the control variables constant.

To illustrate, imagine that Massachusetts passes a new law designed to lower pollution levels at all electronics plants. Connecticut also has many electronics plants, but these plants are not subject to the Massachusetts law. Plants in the two states are very similar except that plants in Massachusetts tend to be larger than plants in Connecticut. A regression of pollution levels on a variable that designates whether the plant is in Massachusetts and on another variable that measures plant size will yield an estimate of the effect of the Massachusetts regulation on pollution levels, holding the size of the plant fixed. If size were to be the only confounder, then this regression would yield a valid estimate of the causal effect of the Massachusetts regulation on pollution levels in electronics plants.

An alternative statistical technique would be to use a matching estimator. For each observation that is subject to the treatment (such as an industrial facility subject to a regulation) the researcher finds a "matching" observation that is not subject to the treatment. To illustrate, let us return to the hypothetical Massachusetts regulation. To implement a matching estimator in this case, the researcher would take each facility in Massachusetts and find a facility in Connecticut of the same size. The researcher would then calculate the difference in pollution levels for the Massachusetts facility and its matching facility in Connecticut. The average of these differences for all Massachusetts plants is the average effect of the regulation on pollution.

Finding a "match" is relatively easy when there is only one confounder (size of the plant in our example). But what if it is important to control not just for size, but also for age of the facility and socioeconomic characteristics of the community, such as the percent employed in manufacturing, population density, median household income, and so forth? To employ a matching estimator in this case, for each facility in Massachusetts the research would need to identify a facility in Connecticut of the same size, age, and with the same socioeconomic characteristics. This may not be possible. This problem is often referred to as the "curse of dimensionality" because the number of dimensions (characteristics) on which facilities must be matched is large. One estimation technique that avoids the curse of dimensionality is matching on the propensity score (Rosenbaum and Rubin, 1983). The propensity score is simply the probability of being treated conditional on the control variables. Observations are then matched on the basis of their propensity to receive treatment, rather than on each individual control variable.

Regression and matching estimates assume that all of the confounders are observable. However, there are frequently cases when there are unobservable factors that are correlated with the treatment as well as potential outcomes. For example, facilities whose managers have a strong personal commitment to the environment may be more likely to participate in certain types of treatment, such as voluntary or so-called "beyond compliance" programs established by government agencies. However, the managers' commitment, which will likely be unobservable to the researcher, is also likely to be correlated with the facility's environmental performance regardless of participation in the program (Coglianese and Nash, 2001). When there are unobservable confounders, standard regression and matching estimators will fail to provide a fully valid estimate of the causal effect of the treatment. In voluntary programs, for example, an ordinary regression estimate will be biased because it will be showing not only the effect of the voluntary program but also the effect of managers' personal commitment to the environment, without being able to separate the level of impact of the two causal factors.

In such cases, alternative estimation strategies need to be used. An estimator known as the *differences-in-differences estimator* can yield a valid estimate of causal effects if the unobservable differences between the treated and nontreated entities are constant over time. For example, imagine that the researcher has data on two sets of facilities: one that participates in a voluntary environmental program and one that does not. However, these two facilities do not have identical indicators of environmental performance before the program is created. In fact, let us suppose that the facilities that participate in the program have, on average, lower pollution levels even before participation. This is depicted graphically in Figure E-3. It is clear from the figure that it would be incorrect to characterize the difference in environmental performance after the program as the causal effect of the regulation, since some of that difference existed before the program came into existence. The differences-in-differences estimator assumes that, in the absence of treatment, the difference in environmental performance would have been the same between the two sets of facilities. The dashed line in Figure E-3 represents the hypothetical pollution levels of the "treatment" plants if they never participated in the program. The causal effect of the program is correctly estimated as the incremental decrease in pollution in the posttreatment period, labeled "treatment effect" in Figure E-3.

Figure E-3 assumes that the unobservable differences are constant over time. If there is good reason to think that they are not, then other estimation strategies will be required. One frequently used estimation technique in such circumstances is the instrumental variables method. To illustrate how this method works, return to the example of a voluntary program where participation is determined, in part, by facility managers' personal commitment to the environment, something that we assume is generally unobservable to the researcher.

For the sake of illustration, imagine that the regulatory agency administering the voluntary program sent letters inviting facilities to participate and did so to a completely *random* sample of facilities. Furthermore, assume that, on average, facilities that received the letter were more likely to

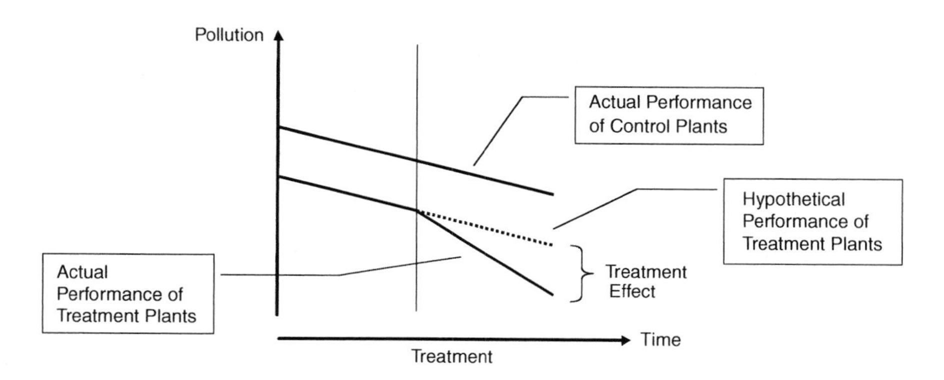

FIGURE E-3 Graphical illustration of the differences-in-differences estimator.

participate than facilities that did not receive the letter; however, some facilities that received the letter did not participate and some facilities that did not receive a letter nonetheless chose to participate. In such a circumstance, the participation decision is not randomly assigned, and traditional statistical estimates of the effect of participation on outcome measures will be biased by the unobservable differences between participants and nonparticipants.

What the instrumental variables estimator does is capitalize on the fact that the government agency randomly assigned facilities to receive the invitation letter. In other words, some set of facilities would participate if they receive a letter and would not participate if they do not receive a letter.[7] For these facilities only, participation is randomly assigned because the letters were randomly sent. The instrumental variables estimator isolates only the effect of participation for those whose participation decisions were determined by whether or not they received a letter.

Although we have highlighted methods only for estimating causal effects, it is clear that these methods are fairly well developed and available for use in evaluating the impacts of environmental policies. These methods, however, have not been widely used in the field of environmental policy, even though they are frequently relied on for evaluation research in other fields. Any effort to increase the role of program evaluation in environmental decision making should therefore seek to encourage research that makes use of these kinds of methods so that reliable inferences can be drawn about the causal effects of environmental policies.

Data Availability and Program Evaluation of Environmental Policies

All of the program evaluation methods we have reviewed here depend on valid and reliable data on environmental outcomes and other nonpolicy determinants of environmental outcomes (such as economic and technological factors). In other fields of policy analysis, researchers have available to them longstanding national surveys such as the Current Population Survey, the National Longitudinal Survey of Youth and the Panel Study of Income Dynamics. For the most part, these kinds of independent longitudinal datasets have not existed for environmental program evaluation.

Much of the data collected on environmental performance are built into the regulations themselves. Thus, researchers have toxics release data available from the TRI, but only on facilities that are subject to the TRI regulations and only for the years during which these regulations have been in effect. Similarly, data are reported by regulated facilities on their air emissions, water discharges, and hazardous waste generation, but these data exist only for the facilities that are regulated under the relevant statutes and for the years in which the regulations have been in effect. This close connection between data and regulation necessarily limits researchers' ability to evaluate the effects of these regulations as a treatment, since the mandated data are not available for unregulated facilities (the control group). However, these data can be used to evaluate the impact of other policies (e.g., voluntary programs, enforcement strategies) by comparing the effects of the treatment on regulated firms subject to the treatment and those not subject to the treatment.

Longitudinal data are available on some ambient environmental conditions (such as air quality), but it is extremely difficult to pinpoint the effects of specific policy changes using these indicators. In most cases it is impossible to use them to identify the effects on individual firms or facilities. Researchers seeking measures of individual firm performance have often used TRI data because they are readily accessible for many, but by no means all, regulated firms. But these data have their limitations too. Most obviously, they do not capture all the impacts firms have on the environment, as the data only cover releases of certain toxic pollutants. Furthermore, these data are self-reported, not adjusted for risk, and reported only by facilities that exceed the established reporting thresholds. All of these factors have been shown to affect the valid use of TRI data as outcome measures for policy evaluation (Graham and Miller, 2001; Snyder, 2004b).[8]

Researchers have sometimes used other measures of environmental impacts, such as biological oxygen demand or total suspended solids levels in water (Gunningham, Kagan, and Thornton, 2003) or levels of water use (Olmstead, Hanemann, and Stavins, 2003). However, obtaining these measures has generally required intensive collection efforts on the part of researchers that have so far limited the use of these data. To a large extent, the future of program evaluation in environmental policy will therefore be married to the future of environmental reporting and performance measurement (Esty, 2001; Metzenbaum, 1998).[9] As we discuss in the next section, this future looks more hopeful than ever before, in part because of new, more uniform and accessible sets of government data on environmental performance.

THE FUTURE OF PROGRAM EVALUATION OF ENVIRONMENTAL POLICY

The idea of subjecting policies to program evaluation research is certainly not new. At about the same time that environmental issues emerged on the federal policy agenda in the 1960s and early 1970s, the federal government also began to emphasize the use of performance evaluations as part of the budgetary process, through efforts such as the Planning, Programming, and Budgeting System, Management by Objective, and Zero-Based Budgeting. These and other attempts to encourage program evaluations of government programs certainly have spilled over into the field of environmental policy from time to time. Yet, compared with other types of government programs, environmental policy has generated only a paucity of systematic program evaluation research. Research funding for environmental program evaluation has lagged behind. Nevertheless, three factors give renewed urgency and ripeness to efforts to expand program evaluation in the field of environmental policy.

First, there are numerous policy innovations that have been implemented in the past 15 years and are now ripe for evaluation. After an initial round of environmental policy making in the early 1970s established the main framework for environmental regulation in the United States, there followed an extended period of concentrated efforts to implement these framework laws. By the late 1980s and early 1990s, however, a variety of factors led to a burst of innovative projects and policies implemented in Washington, D.C., and in the states. This later time period saw the introduction of EPA's "bubble policy," the TRI, and state pollution prevention laws, as well as more recent experimentation with a host of so-called voluntary, public recognition, and regulatory contract programs (such as the EPA's Project XL or the National Environmental Performance Track, or the U.S. Department of Interior's Habitat Conservation Planning program).

Many of these programs have been in place for a sufficiently long time now for their results to be measured through sustained efforts at empirical inquiry. Importantly, many of these programs apply selectively to a subset of all facilities within an industry or sector. Thus, these policies often make it feasible to compare the behavioral responses of participants and nonparticipants (the treatment and control groups). Of course, this does not imply that isolating the causal effect of these policies will be straightforward. The causal effect of voluntary programs is almost always confounded by differences in facilities that explain the decision to participate in the program in the first place—so-called selection effects. But as we discussed above, methods exist to correct for these confounding factors, and some limited evaluation of these programs is already under way that addresses these issues.[10] More research in this area is likely to be highly productive at informing policy makers about the features of different policy initiatives that have been successful, as well as the features that have not been successful at improving environmental outcomes.

Second, the present climate of government management is reasonably conducive to environmental program evaluation. The Government Performance and Results Act (GPRA) requires that all federal agencies devise specific performance goals and report on their achievement of these goals using performance measures. This focus on performance measures, rather than on administrative measures, such as numbers of inspections or numbers of voluntary participants, increases the need for outcome-based evaluation (Sparrow, 2000). Furthermore, the Office of Management and Budget has developed the Performance Assessment Rating Tool (PART) and required that 20 percent of government programs use this tool to evaluate whether they are resulting in significant progress toward public goals (Office of Management and Budget, 2003b). Just as executive orders on the ex ante use of economic analysis for major regulations have given greater prominence to those analytic tools within government agencies, we might expect that GPRA and PART will increase demand within environmental and natural resources agencies for program evaluation research.

Finally, increasing availability and quality of environmental performance data will make it easier for researchers to conduct systematic evaluations of environmental policies. Although the EPA has collected data on air emissions, water discharges, hazardous waste generation, and toxics releases for several decades, in the past these data were collected and maintained separately by the respective program offices within the agency. As a result, each office generated its own metadata and, importantly, its own numbering system for identifying facilities. Thus, the same facility was assigned an Aerometric Information Retrieval System (AIRS) identifier for the air office, a Permit Compliance System identifier for the water office, a TRI identifier for the office of information, and so forth. Researchers hoping to combine data from more than one source were forced to match facilities by hand—usually by name and address. Recently, however, the EPA has instituted a common Facility Registry System identifier. This identifier has been added to all existing EPA databases, allowing researchers to more easily match data on a facility from multiple sources.

Another recent development that is likely to improve environmental policy evaluations is the EPA's Risk-Screening Environmental Indicators (RSEI) model. The RSEI model combines data on toxics releases from the TRI with scientific indicators of the effect of these releases on health risks. By weighting TRI data, RSEI allows researchers to draw inferences about the health effects of policy interventions.

The EPA has also expanded data on regulatory compliance. The Integrated Data for Enforcement Analysis, Enforcement and Compliance History Online, and Online Tracking Information System provide researchers with easier access to certain kinds of data on enforcement and compliance behavior. From 1988 to 2004, EPA went still further in integrating enforcement, compliance, and environmental performance data through the Sector Facility Indexing Project (SFIP). For five industry sectors—automobile assembly, pulp and paper, petroleum refining, iron and steel production, and metal smelting and refining—the SFIP provided one-stop access to data on the number of inspections, compliance with federal regulations, enforcement actions, toxic release levels, and spills. The SFIP database also provided information about the facility, including production capacity and demographic characteristics of the surrounding area.

Although there is much more work to be done to develop and categorize meaningful metrics (Metzenbaum, 2003), recent developments appear headed in a valuable direction for the future of program evaluation research.[11] In Table E-1, we provide information on key types of data currently available for program evaluation of environmental policies. Improvements in data quality and data access, combined with the ripeness of a variety of innovative regulatory instruments and the managerial pressure to evaluate the effectiveness of government programs, suggest that the coming years could be exceptionally promising for program evaluation research on environmental policy.

CONCLUSION

Program evaluation research provides valuable information for policy decision making. The staff and political officials in state and federal regulatory agencies, legislatures, and other oversight bodies (such as the Office of Management and Budget) need to design and implement policies that work to achieve their goals. With information from retrospective evaluations of policies, policy makers will be better able to determine what policies to adopt (and how to design them) in the future. Policy evaluation research can also help identify ways to change existing policies to make them more beneficial.

To be sure, when research shows that policies having intuitive appeal do not yield the anticipated or desired results, some decision makers may remain faithful to their intuitions rather than to what the evidence shows. Resistance to research findings can also occur when actors in the policy process have interests at stake in certain policies. Although these are real considerations, it should be noted that the same was (and still is to a certain extent) the case even in other areas like medicine or education. However, the value of evidence-based practice is only made more compelling when one acknowledges the biases that can otherwise affect decision making.

More program evaluation research should help counteract the skeptical responses to research in the policy process. If a single study demonstrates that a program is effective or ineffective, those who are predisposed to think otherwise may be quick to dismiss the findings. With multiple program evaluation studies on environmental policies, such dismissals will become more difficult to sustain. If several studies reach consistent results, then over time the preponderance of the empirical evidence will be more likely to affect the decisions of policy makers.

Moreover, the reality is that some regulatory officials are receptive to research that can tell them about what works and what does not work. For example, the EPA has recently released a strategy document on environmental management systems that gives priority to the need for careful program evaluation of initiatives in this area (U.S. Environmental Protection Agency, 2004). Both the EPA and the Multi-State Working Group on Environmental Management Systems have sponsored research conferences on management-based strategies for improving environmental performance that have brought together leading researchers from economics and political science (Coglianese and Nash, 2001, 2004).

Only with more efforts to give priority to program evaluation research will decision making over environmental policy be able to become based more on careful deliberation than on rhetorical and political contestation. To be sure, program evaluation research probably will neither end political conflict altogether nor immunize policy makers from all error. But it can help sharpen the focus of policy deliberation as well as inform government's choices about how to allocate scarce resources more effectively. Making program evaluation of environmental policy a priority will be a necessary step toward an evidence-based approach to environmental decision making.

ACKNOWLEDGMENTS

We are grateful for the helpful comments we received from Garry Brewer, Terry Davies, David Heath, Shelley Metzenbaum, Jennifer Nash, Paul Stern, and two anonymous reviewers.

NOTES

1. By the phrase "environmental decision making" we mean to include all policy decisions related to the environment. Although most of the examples throughout this appendix draw on federal pollution-oriented environmental policies in the United States, our discussion applies equally to any type of environmental or natural resources policy decision making at the local, state, federal, and international levels.

2. Risk assessment is not exclusively a scientific enterprise, however, as it often involves making certain policy judgments for which public deliberation may be appropriate (National Research Council, 1996).

3. Sometimes program evaluation researchers distinguish between the "outcomes" and "outputs" of a program. For example, a new enforcement initiative might increase the number of enforcement actions that a regulatory agency brings (an output), but the program evaluation researcher would want to ask whether this new initiative (and the corresponding increase in enforcement actions) actually reduced pollution (an outcome).

4. A comprehensive answer to the question is, of course, beyond the scope of this appendix. For an extensive discussion of the methods of program evaluation research, see Cook and Campbell (1979). King, Keohane, and Verba (1994) also provide a thorough treatment of the methods of qualitative causal inference. Rossi and Freeman (1993) discuss the uses of evaluation methods in the policy process.

5. Rosenbaum (2002) provides a detailed description of a wide range of observational

study designs. Angrist and Krueger (1999) offer an excellent summary of program evaluation methods, as applied to labor policies, including substantially more detail on each of the estimation methods discussed here.

6. When we use the terms "observable" and "unobservable" here, we mean what is observable and unobservable from the perspective of the researcher.

7. In the parlance of the instrumental variables literature, these facilities are labeled compliers. This contrasts with always-takers (facilities that would have participated regardless of whether or not they received the letter), never-takers (facilities that would not have participated regardless of whether they received the letter), and defiers (facilities that would have participated if they did not receive a letter, but would not have participated if they did receive a letter). The instrumental variables method provides a valid estimate of the causal effect of the treatment for compliers (Angrist, Imbens, and Rubin, 1996).

8. A more recent concern is that data may be restricted due to concerns about its potential use by terrorists. For the moment, TRI data continue to be publicly available despite these concerns.

9. In addition to requiring good metrics on outcomes (i.e., environmental performance) for both the treatment and the control groups, policy evaluation also requires data on other potential determinants of environmental performance. These include key variables describing the regulated entities (e.g., production processes, production levels, or market characteristics). Although important work on corporate management has begun to emerge (Andrews, 2003; Prakash, 2000; Reinhardt, 2000), the behavior of firms also remains an area in need of further development.

10. See Alberini and Segerson (2002) for a survey article on evaluation of voluntary programs in the environmental policy area that provides detailed references for evaluations that have addressed issues of selection bias.

11. In addition to developments in the EPA's data management, promising nongovernmental efforts to study and improve different kinds of environmental metrics have also emerged in recent years (O'Malley, Cavender-Bares, and Clark, 2003; Clark, 2002; Esty and Cornelius, 2002; National Academy of Engineering, 1999).

REFERENCES

Alberini, A., and K. Segerson
2002 Assessing voluntary programs to improve environmental quality. *Environmental and Resource Economics* 22:157-184.

Andrews, N.L.
2003 *Environmental Management Systems: Do They Improve Performance?* (Final Report of the National Database on Environmental Management Systems.) Chapel Hill, NC: University of North Carolina.

Angrist, J.D., and A.B. Krueger
1999 Empirical strategies in labor economics. In *The Handbook of Labor Economics, Vol. 3*, O. Ashenfelter, and D. Card, eds. Amsterdam, Holland: Elsevier Science.

Angrist, J.D., G.W. Imbens, and D.B. Rubin
1996 Identification of causal effects using instrumental variables. *Journal of the American Statistical Association* 91(434):444-472.

Arora, S., and T.N. Cason
1995 An experiment in voluntary environmental regulation: Participation in EPA's 33/50 program. *Journal of Environmental Economics and Management* 28(3):271-286.
1996 Why do firms volunteer to exceed environmental regulation? Understanding participation in EPA's 33/50 program. *Land Economics* 72(4):413-432.

Arrow, K.J., M.L. Cropper, G.C. Eads, R.W. Hahn, L.B. Lave, R.G. Noll, P.R. Portney, M. Russell, R. Schmalensee, V.K. Smith, and R.N. Stavins

1996 Is there a role for benefit-cost analysis in environmental, health, and safety regulation? *Science* 272:221-222.

Bardach, E., and R.A. Kagan

1982 *Going by the Book: The Problem of Regulatory Unreasonableness.* Philadelphia, PA: Temple University Press.

Blackman, A., and J. Mazurek

2000 *The Cost of Developing Site-Specific Environmental Regulations: Evidence from EPA's Project XL.* (Discussion Paper 99-35-REV.) Washington, DC: Resources for the Future.

Brewer, G., and P. deLeon

1983 *The Foundations of Policy Analysis.* Homewood, IL: The Dorsey Press.

Bui, L.T.M., and C.J. Mayer

2003 Regulation and capitalization of environmental amenities: Evidence from the toxics release inventory in Massachusetts. *Review of Economics and Statistics* 85(3): 693-708.

Chay, K.Y., and M. Greenstone

2003 *Air Quality, Infant Mortality and the Clean Air Act of 1970.* (NBER Working Paper No. w10053.) Washington, DC: National Bureau of Economic Research.

Clark, W.C.

2002 *The State of the Nation's Ecosystems: Measuring the Lands, Waters, and Living Resources in the United States.* Cambridge, England: Cambridge University Press.

Coglianese, C.

1997 Assessing consensus: The promise and performance of negotiated rulemaking. *Duke Law Journal* 46:1255-1349.

2001 Assessing the advocacy of negotiated rulemaking. *New York University Environmental Law Journal* 9:386-447.

2002 Empirical analysis and administrative law. *University of Illinois Law Review* 2002:1111-1137.

Coglianese, C., and D. Lazer

2003 Management-based regulation: Prescribing private management to achieve public goals. *Law and Society Review* 37(4):691-730.

Coglianese, C., and J. Nash

2001 *Regulating from the Inside: Can Environmental Management Systems Achieve Policy Goals?* Washington, DC: Resources for the Future.

2004 *Leveraging the Private Sector: Management-Based Strategies for Improving Environmental Performance.* (RPP Report No. 6.) Cambridge, MA: Regulatory Policy Program, Center for Business and Government at the John F. Kennedy School of Government, Harvard University.

Coglianese, C., J. Nash, and T. Olmstead

2003 Performance-based regulation: Prospects and limitations in health, safety, and environmental regulation. *Administrative Law Review* 55:741-764.

Cook, T.D., and D.T. Campbell

1979 *Quasi-Experimentation: Design and Analysis Issues for Field Settings.* Boston: Houghton Mifflin Company.

Davies, J.C., and J. Mazurek

1998 *Pollution Control in the United States: Evaluating the System.* Washington, DC: Resources for the Future Press.

de Bruijn, T., and V. Norberg-Bohm

2001 *Voluntary, Collaborative, and Information-Based Policies: Lessons and Next Steps for Environmental and Energy Policy in the United States and Europe.* (RPP Report No. 2.) Cambridge, MA: Regulatory Policy Program, Center for Business and Government at the John F. Kennedy School of Government, Harvard University.

Esty, D.C.

2001 Toward data driven environmentalism: The environmental sustainability index. *Environmental Law Reporter News and Analysis* 31(5):10603-10612.

Esty, D., and P.K. Cornelius, eds.

2002 *Environmental Performance Measurement: The Global Report, 2001-2002.* Oxford, England: Oxford University Press.

Farrow, S.

2000 *Improving Regulatory Performance: Does Executive Office Oversight Matter?* AEI-Brookings Joint Center on Regulatory Studies Working Paper 04-01. Available: http://www.aei.brookings.org/publications/related/oversight.pdf [February 28, 2005].

Graham, M., and C. Miller

2001 Disclosure of toxic releases in the United States. *Environment* 43(8):9-20.

Gunningham, N., R.A. Kagan, and D. Thornton

2003 *Shades of Green: Business, Regulation, and Environment.* Palo Alto, CA: Stanford University Press.

Hahn, R., and P.M. Dudley

2004 *How Well Does the Government Do Cost-Benefit Analysis?* AEI-Brookings Joint Center on Regulatory Studies Working Paper 04-01. Available: http://www.aei-brookings.org/admin/authorpdfs/page.php?id=317 [February 14, 2005].

Hahn, R.W.

1998 Government analysis of the benefits and costs of regulations. *Journal of Economic Perspectives* 12(4):201-210.

Hahn, R.W., and C.R. Sunstein

2002 A new executive order for improving federal regulation? Deeper and wider cost-benefit analysis. *University of Pennsylvania Law Review* 150:1489-1552.

Hamilton, J.T.

1995 Pollution as news: Media and stock market reactions to the toxic release inventory data. *Journal of Environmental Economics and Management* 28(1):98-113.

Hamilton, J.T., and W.K. Viscusi

1999 *Calculating Risks?: The Spatial and Political Dimensions of Hazardous Waste Policy.* Cambridge, MA: MIT Press.

Harrington, W., R.D. Morgenstern, and P. Nelson

2000 On the accuracy of regulatory cost estimates. *Journal of Public Policy Analysis and Management* 19(2):297-322.

Harrington, W., R.D. Morgenstern, and T. Sterner

2004 *Choosing Environmental Policy: Instruments and Outcomes in the United States and Europe.* Washington, DC: Resources for the Future.

Hutter, B.

1989 Variations in regulatory enforcement styles. *Law and Policy* 11(2):153-174.

Khanna, M., and L.A. Damon

1999 EPA's voluntary 33/50 program: Impact on toxic releases and economic performance of firms. *Journal of Environmental Economics and Management* 37(1): 1-25.

Khanna, M., W.R.H. Quimio, and D. Bojilova
1998 Toxics release information: A policy tool for environmental protection. *Journal of Environmental Economics and Management* 36(3):243-266.
King, G., R.O. Keohane, and S. Verba
1994 *Designing Social Inquiry.* Princeton, NJ: Princeton University Press.
Kleindorfer, P., and E. Orts
1998 Informational regulation of environmental risks. *Risk Analysis* 18:155-170.
Konar, S., and M.A. Cohen
1997 Information as regulation: The effect of community right to know laws on toxic emissions. *Journal of Environmental Economics and Management* 32(2):109-124.
Langbein, L., and C. Kerwin
2000 Regulatory negotiation versus conventional rule making: Claims, counterclaims, and empirical evidence. *Journal of Public Administration Research and Theory* 10:599-632.
Marcus, A., D.A. Geffen, and K. Sexton
2002 *Reinventing Environmental Regulation: Lessons from Project XL.* Washington, DC: Resources for the Future.
May, P., and S. Winter
2000 Reconsidering styles of regulatory enforcement: Patterns in Danish agro-environmental inspection. *Law and Policy* 22:143-173.
Metzenbaum, S.
1998 *Making Measurement Matter: The Challenge and Promise of Building a Performance-Focused Environmental Protection System.* (Report No. CPM-92-2.) Washington, DC: Brookings Institution Center for Public Management.
2003 More nutritious beans. *Environmental Forum* April/May:18-41.
Morgenstern, R., ed.
1997 *Economic Analyses at EPA: Assessing Regulatory Impact.* Washington, DC: Resources for the Future.
National Academy of Engineering
1999 *Industrial Environmental Performance Metrics. Challenges and Opportunities.* Committee on Industrial Environmental Performance Metrics. Washington, DC: National Academy Press.
National Research Council
1983 *Risk Assessment in Federal Government: Managing the Process.* Committee on the Institutional Means for Assessment of Risks to Public Health. Commission on Life Sciences. Washington, DC: National Academy Press.
1996 *Understanding Risk: Informing Decisions in a Democratic Society.* P.C. Stern and H.V. Fineberg, eds. Committee on Risk Characterization. Commission on Behavioral and Social Sciences and Education. Washington, DC: National Academy Press.
Office of Management and Budget, Office of Information and Regulatory Affairs
2003a *Circular A-4: Regulatory Analysis.* Available: http://www.whitehouse.gov/omb/circulars/a004/a-4.pdf [February 14, 2005].
2003b Testimony of the Honorable Clay Johnson III, Deputy Director for Management, before the Committee on Government Reform, U.S. House of Representatives, September 18, 2003. Available: http://www.whitehouse.gov/omb/legislative/testimony/cjohnson/030918_cjohnson.html [February 14, 2005].
Olmstead, S.M., W.M. Hanemann, and R.N. Stavins
2003 *Does Price Structure Matter?: Household Water Demand Under Increasing-Block and Uniform Prices.* New Haven, CT: School of Forestry and Environmental Studies, Yale University.

O'Malley, R., K. Cavender-Bares, and W.C. Clark

2003 Providing "better" data: Not as simple as it might seem. *Environment* 45(May): 8-18.

Prakash, A.

2000 *Greening the Firm: The Politics of Corporate Environmentalism.* Cambridge, England: Cambridge University Press.

Reinhardt, F.

2000 *Down to Earth: Applying Business Principles to Environmental Management.* Boston: Harvard Business School Press.

Revesz, R.L., and R.B. Stewart, eds.

1995 *Analyzing Superfund: Economics, Science and Law.* Washington, DC: Resources for the Future.

Rosenbaum, P.R.

2002 *Observational Studies.* New York: Springer-Verlag.

Rosenbaum, P.R., and D.B. Rubin

1983 The central role of the propensity score in observational studies for causal effects. *Biometrika* 70(1):41-55.

Rossi, P.H., and H.E. Freeman

1993 *Evaluation: A Systematic Approach*, Fifth Edition. Newbury Park, CA: Sage Publications.

Scholz, J.T.

1984 Cooperation, deterrence and the ecology of regulatory enforcement. *Law & Society Review* 18:179-224.

Shimshack, J.P., and M.B. Ward

2003 *Enforcement and Environmental Compliance: A Statistical Analysis of the Pulp and Paper Industry.* Medford, MA: Tufts University.

Snyder, L.D.

2004a Are management-based regulations effective? Evidence from state pollution prevention programs. In *Essays on Facility Level Response to Environmental Regulation.* Ph.D. dissertation. Cambridge, MA: Harvard University.

2004b Are the TRI Data valid measures of facility-level environmental performance?: Reporting thresholds and truncation bias. In *Essays on Facility Level Response to Environmental Regulation.* Ph.D. dissertation. Cambridge, MA: Harvard University.

Sparrow, M.

2000 *The Regulatory Craft: Controlling Risks, Solving Problems, Managing Compliance.* Washington, DC: Brookings Institution Press.

Stavins, R.N.

1998 What can we learn from the grand policy experiment? Positive and normative lessons from SO_2 allowance trading. *Journal of Economic Perspectives* 12(3): 69-88.

2003 Experience with market-based environmental policy instruments. In *Handbook of Environmental Economics, Volume 1 Environmental Degradation and Institutional Responses,* K.-G. Maler, and J.R. Vincent, eds. Amsterdam, Holland: North-Holland Press.

Stokey, E., and R. Zeckhauser

1978 Project evaluation: Benefit-cost analysis. In *A Primer for Policy Analysis.* New York: W.W. Norton & Company.

U.S. Environmental Protection Agency, EMS Permits and Regulations Workgroup

2004 *EPA's Strategy for Determining the Role of Environmental Management Systems in Regulatory Programs.* Washington, DC: U.S. Environmental Protection Agency.

U.S. Environmental Protection Agency, Office of Air and Radiation
1997 *Final Report to Congress on Benefits and Costs of the Clean Air Act, 1970 to 1990.* (EPA 410-R-97-002.) Washington, DC: U.S. Environmental Protection Agency.
U.S. Environmental Protection Agency, Office of the Administrator
2000 *Guidelines for Preparing Economic Analyses.* (EPA 240-R-00-003.) Washington, DC: U.S. Environmental Protection Agency.
Viscusi, W.K.
1996 Regulating the regulators. *University of Chicago Law Review* 63:1423-1461.

TABLE E-1 Data Sources for Program Evaluation of Environmental Policy

Topic	Data	Source	Description	Types of Facilities Covered
Data on Outcomes				
Toxics and Hazardous Waste	Toxics Release Inventory	Self-reported by facilities	Contains data on pounds of chemicals released to air, water, land, underground injection, and transferred off site. Also includes data on pollution prevention activities and recycling.	Manufacturing facilities that meet certain thresholds.
	Comprehensive Environmental Response, Compensation and Liability Information System (CERCLIS)		Contains data on Superfund sites, including whether they are on the National Priority List, ownership information, dates and descriptions of actions taken.	Superfund sites
	Record of Decisions		Provides *.pdf files of decisions regarding Superfund sites.	Superfund sites

	Resource Conservation and Recovery Act Information (RCRAInfo)		Contains data on hazardous waste generation for large quantity generators of hazardous waste and disposal information for all treatment, storage and disposal facilities. Replaces two previously maintained databases, the Biennial Reporting System and the Resource Conservation and Recovery Information System.	Generators of hazardous waste and hazardous waste treatment storage and disposal facilities
Water	Permit Compliance System (PCS)	Discharge data are self-reported by facilities. Other information entered and maintained by either the EPA or the states.	Contains data on permit limits, discharge levels, enforcement, and inspection activities.	All National Permit Discharge and Elimination System permit holders.
	Safe Drinking Water Information System	Maintained by the EPA or designated states.	Contains data on drinking water contaminant violations and enforcement actions.	Public drinking water systems
Air	Aerometric Information Retrieval System (AIRS) Facility Subsystem	Self-reported by facilities.	Contains data on permits, emissions, inspection, and compliance with air quality standards.	All air permit holders

Continued

TABLE E-1 Continued

Topic	Data	Source	Description	Types of Facilities Covered
Compliance and Enforcement	Enforcement and Compliance History Online	Combined enforcement and compliance data from PCS, AIRS, and RCRAInfo.	Contains data on inspection and compliance for water, air, and hazardous waste permit holders.	Same as underlying PCS, AIRS, and RCRAInfo databases.
	Integrated Data for Enforcement Analysis	Combined enforcement and compliance data from PCS, AIRS, and RCRAInfo.	Contains data on inspection and compliance for water, air, and hazardous waste permit holders.	Same as underlying PCS, AIRS, and RCRAInfo databases.
Data on Covariates				
Firm Data	Compustat	Standard and Poor's	Contains income, balance sheet, and cash flow data.	Publicly held companies. Data are available by subscription.

	Dunn and Bradstreet Million Dollar Database	Dunn and Bradstreet	Contains data on sales, employment, industry, and ownership.	1.6 million U.S. and Canadian companies, both private and public. Data are proprietary and available by subscription only.
Plant data	Dunn and Bradstreet Million Dollar Database	Dunn and Bradstreet	Contains employment information at plant and firm level.	1.6 million U.S. and Canadian companies, both private and public. Data are proprietary and available by subscription only.
	Longitudinal Research Database	U.S. Census Bureau	Contains data from the Census of Manufacturers and the Annual Survey of Manufacturers. Data include employment, product classes, and shipments.	Available only by approved proposal at one of eight regional data centers.

Appendix F

Panel Members, Staff, and Contributors

PANEL MEMBERS AND STAFF

GARRY BREWER *(Chair)* is the Frederick K. Weyerhaeuser professor of resource policy and management at Yale University. In the field of policy science, his expertise involves environmental management. He was first appointed to the faculty of the School of Management in 1974. In 1980 he joined the faculty of the School of Forestry & Environmental Studies, and became the first Frederick K. Weyerhaeuser chair (1984 to 1990). He also occupied the Edwin W. Davis Chair from 1990 to 1991. Brewer has served as dean and professor of the University of Michigan's School of Natural Resources & Environment, professor at the Michigan Business School, and as dean and member of the faculty at the University of California at Berkeley. He has served on and chaired numerous national and international panels and commissions, including those of the National Research Council (NRC), the International Institute for Applied Systems Analysis, the Department of Energy, the Nuclear Waste Technical Review Board, the American Association for the Advancement of Science, and Sweden's National Foundation for Strategic Environmental Research. A graduate of the University California at Berkeley, he has an M.S. in public administration from San Diego State University and M.Phil. (1968) and Ph.D. (1970) degrees from Yale in political science.

BRADEN R. ALLENBY is professor of civil and environmental engineering, and of law, at Arizona State University, as well as a Batten fellow in residence at the University of Virginia's Darden Graduate School of Business Administration. Previously he was the environment, health and safety vice president of AT&T and served as director for energy and environmental systems at the Lawrence Livermore National Laboratory. He is a member of the Virginia Bar and has worked as an attorney for the Civil Aeronautics Board and the Federal Communications Commission; he has also worked as a strategic consultant on economic and technical telecommunications issues. His publications include *Design for Environment* (1997), *Industrial Ecology* (2003), *and Industrial Ecology: Policy Framework and Implementation* (1999). He writes a column for the *Green Business Letter* and is coeditor of *The Greening of Industrial Ecosystems* (1994) and *Environmental Threats and National Security* (1994). He has taught courses at the Yale University School of Forestry and Environmental Studies, Princeton Theological Seminary, Columbia University, and the University of Virginia School of Engineering. A cum laude graduate of Yale University in 1972, Allenby received his Juris Doctor from the University of Virginia Law School in 1978 and his master's in economics from the University of Virginia in 1979. He received his master's in environmental sciences from Rutgers University in 1989 and his Ph.D. in environmental sciences from Rutgers in 1992.

RICHARD ANDREWS is the Thomas Willis Lambeth distinguished professor of public policy at the University of North Carolina, Chapel Hill, with appointments in the Departments of Public Policy, Environmental Sciences and Engineering, and City and Regional Planning, the Carolina Environmental Program, and the Curriculum in Ecology. His research and teaching focus on environmental policy in the United States and worldwide; he is the author of *Managing the Environment, Managing Ourselves* (1999) and *Environmental Policy and Administrative Change* (1976); he has conducted research projects on environmental policy innovations in the United States, the Czech Republic, and Thailand. His recent research addresses the effects of public policies as incentives for environmental decisions by businesses, particularly "voluntary" approaches such as self-regulation by corporate and business customer mandates for introduction and third-party auditing of environmental management systems. He has chaired or served on study committees for the NRC, the Science Advisory Board of the Environmental Protection Agency, the National Academy of Public Administration, and the Office of Technology Assessment. Before joining the Carolina faculty in 1981, he taught at the University of Michigan's School of Natural Resources and was a Peace Corps volunteer in Nepal. He has an undergraduate degree from Yale University and a Ph.D. and a professional master's degree from the University of North Carolina's Department of City and Regional Planning.

SUSAN CUTTER is a Carolina distinguished professor of geography at the University of South Carolina. She is also the director of the Hazards Research Lab, a research and training center that integrates geographical information processing techniques with hazards analysis and management. She is the cofounding editor of an interdisciplinary journal, *Environmental Hazards*. She has been working in the risk and hazards fields for more than 25 years and has authored or edited numerous books and peer-reviewed articles. She coauthored *The Geographical Dimensions of Terrorism* (2003) and *Exploitation, Conservation, Preservation: A Geographic Perspective on Natural Resource Use* (2003). She also authored *American Hazardscapes: The Regionalization of Hazards and Disasters* (2001), which chronicles the increasing hazard vulnerability to natural disaster events in the United States during the past 30 years. She was elected a fellow of the American Association for the Advancement of Science and served as president of the Association of American Geographers in 1999-2000. She received her Ph.D. from the University of Chicago in 1976.

J. CLARENCE DAVIES is a senior fellow in the Risk, Resource, and Environmental Management division at Resources for the Future. Previously he was the assistant administrator for policy, planning and evaluation of the U.S. Environmental Protection Agency (EPA). In that position, he had responsibility for oversight of all agency policies and programs, as well as the regulatory process. Davies was coauthor of the original plan that created EPA and has long served as an adviser to the agency. He also served with the Council on Environmental Quality and the Bureau of the Budget (now the Office of Management and Budget). He has been executive vice president of the Conservation Foundation, where he specialized in matters concerning toxics, risk assessment, and the control and integrated management of pollution, and has served as a senior staff member at the Council on Environmental Quality. He chaired the NRC's Committee on Principles of Decision Making for Regulating Chemicals in the Environment. He has been on the faculty of Princeton University and Bowdoin College. He is the author of *The Politics of Pollution* and *Neighborhood Groups and Urban Renewal* and the coauthor or coeditor of numerous other books, articles, and monographs dealing with environmental issues. Davies has a B.A. in American government from Dartmouth College and a Ph.D. in American government from Columbia University.

LOREN LUTZENHISER is professor of urban studies and planning at Portland State University, where he also serves as director of the urban studies Ph.D. program. His teaching interests include environmental policy and practice, particularly in terms of energy infrastructures and technological change, urban environmental sustainability, and the built environment.

His research focuses on the environmental impacts of sociotechnical systems, particularly how urban energy/resource use is linked to global environmental change. Recent studies have considered variations across households in energy consumption practices, how energy-using goods are procured by government agencies, how commercial real estate markets work to develop both poorly performing and environmentally exceptional buildings, and how the "greening" of business may (or may not) be influenced by local sustainability movements and business actors. He is currently completing a major study for the State of California on the behavior of households, businesses and governments in the aftermath of that state's 2001 energy deregulation crisis. Lutzenhiser has published widely in social science, policy, and applied journals. Prior to entering academia, he worked as a local antipoverty program director and regional social program planner. He is a member of the editorial board of *Social Problems* and the past chair of the American Sociological Association's section on environment and technology. He has B.A. and M.A. degrees in sociology from the University of Montana and a Ph.D. in sociology from the University of California, Davis.

BONNIE McCAY is a Board of Governors distinguished service professor of anthropology and ecology at Cook College, Rutgers State University. She has been with the university since 1974. In that time she has served in many roles, including director of the Center for Environmental Indicators and chair of the Department of Human Ecology. Her research and publications focus on the social implications of fishery management, user group participation in regulatory processes, the culture of fishing communities, fishery comanagement, and the socioeconomic consequences of various property rights regimes in fishery resources. She received the Norwegian Marshall Fund Award for Research in Marine Conservation and is a fellow of the American Association for the Advancement of Science. She is author, coauthor, or editor for numerous books, the most recent being *Enclosing the Commons: Individual Transferable Quotas in a Nova Scotia Industry* (2002). She was a member of the Ocean Studies Board of the NRC and is a national associate of the NRC, having served on several other NRC committees. She is vice-chair of the Federal Advisory Committee on Marine Protected Areas and serves on the Panel on Development of Wind Turbine Facilities in Coastal Waters for the State of New Jersey. She has a B.A. from Portland State University and M.Phil. and Ph.D. degrees, the latter in anthropology, from Columbia University.

TIMOTHY McDANIELS is a professor at the University of British Columbia, where he teaches in two graduate programs: the School of Community and Regional Planning and the Institute for Resources, Environment and

Sustainability. He is the interim director of the institute and also directs the eco-risk research unit at the university. He is a specialist in decision analysis, value elicitation, policy analysis, and the social dimensions of risk management questions. He enjoys working on ecological, technology, and human health risk issues. He has a special interest in the underlying concepts, design and implementation of stakeholder-based, structured decision processes for risk management questions. He is also an adjunct professor in engineering and public policy at Carnegie Mellon University, where he is a coinvestigator in the new Climate Decision Making Center. He has served as a member of national peer review and advisory committees for the Environmental Protection Agency, the National Oceanographic and Atmospheric Administration, and Health Canada. He is the decision sciences editor for *Risk Analysis*, and has served as editor for *Risk, Decision and Policy*. He is a fellow of the Society for Risk Analysis and has won the society's distinguished service award. He is coeditor of *Risk Analysis and Society: An Interdisciplinary Characterization of the Field* (2003). He received his B.A. in economics in 1970 from the University of Minnesota, an M.A. in economics in 1974 from Simon Fraser University, and Ph.D. in decision sciences and policy analysis in 1989 from Carnegie Mellon University.

JENNIFER NASH is director of the Regulatory Policy Program at the Center for Business and Government, John F. Kennedy School of Government, Harvard University. She conducts empirical research on emerging trends in government regulation and industry self-regulation; current research explores the effectiveness of performance- and management-based regulation in achieving policy goals and the role of voluntary programs in improving the environmental performance of firms. She is coeditor of *Regulating From the Inside: Can Environmental Management Systems Achieve Policy Goals?* (2001) and the forthcoming *Leveraging the Private Sector: Management-Based Strategies for Improving Environmental Performance*. Before coming to the Kennedy School, she served as associate director and acting director of the Technology, Business, and Environment Program at the Massachusetts Institute of Technology, as a planner for the Massachusetts Division of Capital Planning and Operations, and as executive director of the Delaware Valley Citizen's Council for Clean Air. She has a master of city planning degree from the Massachusetts Institute of Technology (1988).

STEVEN W. PERCY, adjunct lecturer of corporate strategy and international business at the University of Michigan Business School, is the former chair and chief executive officer of BP America, Inc., which was BP's U.S. subsidiary prior to its merger with Amoco Corporation. Prior to assuming those duties, he was president of BP Oil in the United States. Since retiring from BP, he has served as the head of Phillips Petroleum's Refining, Marketing and Transportation Company. He has been a senior planning analyst with Babcock & Wilcox and in several managerial positions with Standard Oil before its merger with BP. He is a member of the board of directors of Omnova Solutions Inc., Wavefront Energy and Environmental Services, Inc., and Resources for the Future. He served as a member of President Clinton's Council on Sustainable Development in the role of cochair of its climate change task force. He has chaired Cleveland State University's Foundation, held the position of vice chair of the Greater Cleveland Growth Association and chair of its Finance Committee, and chair of Neighborhood Progress, Inc. Percy has a B.S. in mechanical engineering from Rensselaer Polytechnic Institute, an M.B.A. from the University of Michigan, and a J.D. degree from Cleveland Marshall College of Law. He is a member of the Ohio State Bar.

DAVID SKOLE is a professor and director of the Center for Global Change and Earth Observations, a research program focused on environmental research using remote sensing systems at Michigan State University. His research interests focus on the role humans play in changing land cover throughout the world. He uses satellite data to measure the patterns of landscape change at regional and global scales and employs field research to uncover the fundamental processes of change. He is also developing analyses and models of the carbon cycle and biodiversity. Currently he is involved in research projects focused on understanding the interannual variation in deforestation rates, as well as the social and ecological controls on its variation over time. He is a principal investigator on several funded research projects, including the Tropical Rain Forest Information Center, the Large Scale Amazon Basin Experiment, the Landsat 7 Science Team, the Canadian Radarsat program, and the Japanese JERS program. He has served on several committees of the National Aeronautical and Space Administration and panels for the Earth Observing System and its data system and other programs. He is currently the high resolution design team leader for the United Nations project on Global Observations of Forest Cover of the Committee on Earth Observation Satellites. He has an M.S. in environmental science from Indiana University and a Ph.D. in natural resources from the University of New Hampshire.

PAUL C. STERN *(Study Director)* is also study director of two NRC committees: the Committee on the Human Dimensions of Global Change and the Committee on Assessing Behavioral and Social Science Research on Aging. His research interests include the determinants of environmentally significant behavior, particularly at the individual level, participatory processes for informing environmental decision making, and the governance of environmental resources and risks. He is the coauthor or coeditor of *Envi-*

ronmental Problems and Human Behavior (2002), *The Drama of the Commons* (2002), and *New Tools for Environmental Protection: Education, Information, and Voluntary Measures* (2002). Stern is a fellow of the American Association for the Advancement of Science and the American Psychological Association. He has a B.A. from Amherst College and M.A. and Ph.D. degrees from Clark University.

CONTRIBUTORS

WILLIAM ASCHER is the Donald C. McKenna professor of government and economics at Claremont McKenna College, where he also serves as vice president and dean of the faculty. His work on decision making and environmental/natural resource issues includes *Forecasting: An Appraisal for Policymakers and Planners; Strategic Planning and Forecasting; Natural Resource Policymaking in Developing Countries;* and *Why Governments Waste Natural Resources*. He currently serves on the Environmental Protection Agency's Scientific Advisory Board Committee on Valuing the Protection of Ecological Systems and Services.

LORI SNYDER BENNEAR is assistant professor of environmental economics and policy at the Nicholas School of the Environment and Earth Sciences at Duke University. Her research focuses on estimating the effect of different regulatory innovations on measures of facility-level environmental performance, such as pollution levels, chemical use, and technology choice. Her recent work has focused on measuring the effectiveness of management-based regulations, which require each regulated entity to develop its own internal rules and initiates to achieve reductions in pollution, as well as the effectiveness of regulations that mandate public reporting of toxic emissions.

CARY COGLIANESE is visiting professor of law at the University of Pennsylvania Law School and associate professor of public policy and chair of the Regulatory Policy Program at the John F. Kennedy School of Government at Harvard University. His interdisciplinary research focuses on regulatory policy and administrative law, with a particular emphasis on the empirical evaluation of alternative and innovative regulatory strategies. His recent work has focused on the application of management-based and performance-based regulation to health, safety, and environmental problems; the role of science, economics, and information in the regulatory process; and the effects of consensus-building on regulatory policy making.

ROBIN GREGORY is senior researcher with Decision Research in Vancouver, British Columbia (head office: Eugene, Oregon). His research and applied consulting is focused on topics related to facilitating meaningful stakeholder participation as part of environmental policy deliberations, encouraging learning and adaptive resource management, using decision-aiding techniques to evaluate nonmarket benefits and costs, and understanding processes of preference construction and elicitation.

ANDREW J. HOFFMAN is the Holcim professor of sustainable enterprise at the University of Michigan, a position with joint appointments at the Stephen M. Ross School of Business and the School of Natural Resources and Environment. In this role, he also serves as the faculty codirector of Michigan's dual degree (MS/MBA) Corporate Environmental Management Program. His research focuses on the cultural and institutional aspects of corporate environmental and sustainability strategies. He is the author or editor of four books and has been the recipient of the Rachel Carson Prize (from the Society for Social Studies of Science) and the Rising Star Award (from the World Resources Institute/Aspen Institute).

REBECCA J. ROMSDAHL worked with the committee through the National Academies graduate fellowship program. She is a doctoral candidate in environmental science and public policy at George Mason University. Her dissertation work examines the Federal Advisory Committee Act of 1972 and the challenges it presents for deliberative public participation in natural resource management decisions.